Research on Properties of Cement-Based Materials and Concrete

Research on Properties of Cement-Based Materials and Concrete

Editors

Dongyi Lei
Ying Li
Guo Yang
Jingbin Yang

MDPI

Basel • Beijing • Wuhan • Barcelona • Belgrade • Novi Sad • Cluj • Manchester

Editors

Dongyi Lei
Qingdao University of Technology
Qingdao
China

Ying Li
Qingdao University of Technology
Qingdao
China

Guo Yang
Hohai University
Nanjing
China

Jingbin Yang
Tongji University
Shanghai
China

Editorial Office
MDPI AG
Grosspeteranlage 5
4052 Basel, Switzerland

This is a reprint of articles from the Special Issue published online in the open access journal *Buildings* (ISSN 2075-5309) (available at: https://www.mdpi.com/journal/buildings/special_issues/HVYSNK8FQB).

For citation purposes, cite each article independently as indicated on the article page online and as indicated below:

Lastname, A.A.; Lastname, B.B. Article Title. *Journal Name* **Year**, *Volume Number*, Page Range.

ISBN 978-3-7258-2575-2 (Hbk)
ISBN 978-3-7258-2576-9 (PDF)
doi.org/10.3390/books978-3-7258-2576-9

Contents

buildings

Article

Study on the Performance of Polymer-Modified Conductive Cement-Based Materials

Min Li [1,2], Jianjun Zhong [1,*], Guoqing Li [1], Qianyi Zhang [1], Feng Cen [1] and Peiwei Gao [1,*]

[1] College of Civil Aviation, Nanjing University of Aeronautics and Astronautics, Nanjing 211106, China; bt1601013@nuaa.edu.cn (M.L.); 18831481808@163.com (G.L.); zoey.j@nuaa.edu.cn (Q.Z.); 18862326148@163.com (F.C.)

[2] School of Road and Bridge Engineering, Xinjiang Vocational and Technical College of Communications, Urumqi 831401, China

* Correspondence: zhongjianjun@nuaa.edu.cn (J.Z.); gpw1963@nuaa.edu.cn (P.G.)

Abstract: In order to study the synergistic effect of polymer and conductive functional materials on the properties of cement-based materials, polymer conductive cement-based materials were prepared by mixing four polymer lotions of silicon–acrylate emulsion (SG), phenylacrylic emulsion (SR), waterborne epoxy resin emulsion (SH), and acrylic emulsion (SX) with carbon fiber (CF) and carbon black (CB), two conductive functional materials, in a certain proportion. The effects of the different polymer–cement ratios (P/C) of the four polymers on the physical, mechanical, and electrical properties of conductive cement-based materials were studied. The results illustrated that SH improved the fluidity of cement paste, and the four polymers all had a delaying effect, which led to the hardening of the specimens and the extension of the demoulding specimens to varying degrees. SH and SR can increase the ratio of flexural strength to compressive strength (F/C) in cement paste and improve the toughness of materials, and the maximum value is reached when the P/C is 0.15. Except for SX, the other three polymer lotions can reduce the resistivity of cement paste, which is beneficial to the improvement of conductivity. The improvement sequence is SH > SR > SG. Among them, both SH group and SR group achieved the lowest electrical resistivity at the P/C of 0.15. The four kinds of polymer lotion can significantly reduce the water absorption of the specimen and promote the waterproof performance. The improvement effect: SH > SR > SG > SX. Among them, both the SH group and SR group achieved the minimum water absorption at the P/C of 0.15.

Keywords: polymer; carbon fiber (CF); carbon black (CB); polymer–cement ratio (P/C); liquidity; flexural strength to compressive strength (F/C); resistivity

Citation: Li, M.; Zhong, J.; Li, G.; Zhang, Q.; Cen, F.; Gao, P. Study on the Performance of Polymer-Modified Conductive Cement-Based Materials. *Buildings* **2023**, *13*, 2961. https://doi.org/10.3390/buildings13122961

Academic Editor: Jan Fořt

Received: 21 October 2023
Revised: 10 November 2023
Accepted: 23 November 2023
Published: 28 November 2023

1. Introduction

Polymer cement-based materials are composite materials made by adding polymers as modifiers or reinforcing agents to conventional cement pastes [1]. The proportion of polymers in the matrix is expressed by the polymer–cement ratio (polymer/cement). Conductive polymer cement-based materials are made of cementitious materials such as cement and fly ash, and raw materials such as conductive fillers, polymers, and coarse and fine aggregates.

Common conductive components include carbon fiber [2–4], graphene [5], carbon nanotube [6–8], carbon black [9], graphite [10], steel fiber [11], etc. After single or compound doping and surface treatment in different forms, excellent conductive effect will be formed in the matrix and applied to the melting of ice and snow in roads and bridges [12,13] as well as building heating [14]. Due to the excellent flexibility and viscoelastic behavior of polymers, they are widely used in cement-based materials to improve the brittle-type structure of cement-based materials [15]. Polymers that can promote the performance of cement-based composite materials mainly include styrene–butadiene emulsion, polyurethane, epoxy resin, hydroxypropyl methyl cellulose ether, polyacrylamide, etc.

CF is a kind of high-performance carbon fiber material, which has the advantages of good conductivity, electrothermal property, high temperature resistance, corrosion resistance, and oxidation resistance. Cement products with conductive function can be prepared by adding CF into cement-based materials in a certain proportion. CF can be divided into two categories: one is PAN-based CF made from polyacrylonitrile, and the other is pitch-based CF made from pitch. Among them, the price of PAN-based CF is 5~10 times that of pitch-based CF [16]. The factors affecting the use effect of CF are the length, content, and dispersion. CF has different lengths, such as 3 mm, 6 mm, 9 mm, and carbon nano fibers. Shuhua Liu et al. [17] studied the workability, mechanical properties, electrical conductivity, and electrothermal effect of CF-modified conductive concrete (ECON). The results show that ECON, with a compressive strength of 30 MPa and a resistivity of 0.1 Ω·cm, can be prepared by using CF with volume content of 0.75% and length of 9 mm. Ilhwan You et al. [18] studied the effects of CF length, electrode spacing, and probe configuration on the conductivity of cement-based composites. The results show that when designing electrodes for cement composites containing CF, the CF length should be shorter than the electrode spacing. Ting Ding et al.'s [19] study found that CF content had a significant effect on resistivity, and that when the carbon fiber content was 0.65%, the penetration threshold was reached, and the resistivity of CF conductive concrete tended to be stable. CF content is reduced within a certain range, and the workability of carbon fiber conductive concrete is improved. Iftekar Gull et al. [20] observed that in surfactant-mixed water, a lower-speed mechanical stirring device successfully deagglomerates CF filaments without interfering with their individual morphology, resulting in a better distribution of fibers in concrete.

Although CF has good electrical conductivity, it is expensive and not suitable for a large number of applications. Single-phase conductive concrete is limited by the nature of the material, and there are more or less shortcomings in all aspects. The composite mixing of two or more conductive phase materials can simultaneously exert the characteristics of several materials, play a role of mutual compensation and mutual promotion, and make it is easier to obtain conductive concrete with excellent performance, cost, and other aspects. The use of CB instead of CF can reduce costs and improve performance under the premise of ensuring the function of the material. Dehghanpour et al. [21] prepared 36 different kinds of concrete containing CF, recycled RNCB, waste wire erosion, and steel fiber. The test results show that RNCB has a significant effect on reducing resistance when used in combination with other fillers. Dong et al. [22] developed conductive rubber fiber-filled CB cement-based composites. The conductivity of 0.5% CB filled composites increased with the increase in rubber content. The composites with a CB content < 4.0% had better compressive strength. In this paper, based on the analysis of a large number of existing research findings, combined with the existing research results of the research group, the commonly used materials CF and CB were selected as conductive materials to prepare conductive cement-based composites, and polymer modification research was further carried out.

The addition of polymer can promote the durability of cement materials, such as impermeability, chemical-corrosion resistance, and adhesion between aggregate and cement materials. It can also reduce later shrinkage rate and improve mechanical properties [23]. Long et al. [24] found that the macromolecular structure formed by the reaction between acrylic emulsion and cement hydration product $Ca(OH)_2$, and the combination of ionic bonds, improves the structural compactness and effectively enhances the impermeability of the material. Ohama et al. [23] proposed the following modification mechanism of polymer in cement matrix. After the polymer is added, the C-S-H gel is slowly formed with the hydration of the cement, and the polymer particles accumulate on the surface of the cement particles and the C-S-H gel. With the gradual completion of hydration and the hardening of the slurry, the polymer particles are limited in the capillary pores and flocculate. A layer of polymer film is formed on the surface of the particles, filling some large voids. Finally, with the evaporation of water and the completion of hydration, the polymer and cement hydration products are bonded to each other, which improves the compactness and the microscopic pore structure of the material.

Schulze et al. [25] found that the macroscopic properties of styrene–butadiene-modified mortar are greatly affected by water–cement ratio and curing conditions. Wang Peiming et al. [26] found that the addition of styrene butadiene rubber emulsion can reduce water consumption, improve the mechanical characteristics of mortar, such as bending strength and binding strength, and improve the shrinkage deformation performance and durability of materials. Bureau et al. [27] carried out a three-point bending test on styrene–butadiene rubber emulsion-modified mortar. When the water–cement ratio was 0.45, the flexural strength of the mortar enlarged with the increase in styrene–butadiene rubber emulsion content. However, when the content was more than 10%, although the flexural strength continued to increase, the compressive strength decreased [28]. Yang et al. [29] explored the effect of waterborne epoxy and lignin epoxy on the mechanical properties of polymer-modified cement mortar based on the effective emulsification of waterborne epoxy with active emulsifier. It was found that the toughening effect of epoxy on mortar was obvious. With the increase in P/C, the flexural strength of the specimens increased and the compressive strength decreased. Aggarwal et al. [30] found that the waterproof performance of concrete was improved with the incorporation of acrylic polymer, which had the conditions for application in humid environment. With the continuous progress of research, polymer composite materials in different forms are widely used in concrete waterproof, anti-corrosion, and repair engineering field [14]. Based on its excellent performance, this paper intends to add polymers on the basis of conductive cement-based materials to study its modification mechanism.

So far, there are many studies on the properties of polymer-modified cementitious materials, and a large number of research results have been produced [31]. However, there are a few pieces of research on polymer-modified conductive cement-based materials. In this paper, from the perspective of improving the toughness, electrothermal performance, and durability of cement-based materials, four polymers of SG, SR, SH, SX, and two conductive functional materials of CF and CB were selected to carry out the preparation and performance research of polymer conductive cement-based materials, so as to select the best polymer to be applied to conductive cement-based materials and improve their physical, mechanical, and electrothermal properties.

2. Materials and Methods

2.1. Raw Materials

The cement was P.O 42.5 grade cement produced by a factory in Nanjing, Jiangsu Province, and the silica fume was micro silica fume produced by a factory in Nanjing, Jiangsu Province. Admixtures included polycarboxylate superplasticizer. The water reduction rate was about 30–40%. The methyl cellulose dispersant was a white granular powder. Tributyl phosphate was used as a defoaming agent, and magnesia expansive agent was mainly composed of MgO and CaO.

Polymer emulsion: SH, solid content (50 ± 1)%. SR, solid content (45 ± 1)%. SG, solid content (48 ± 1)%. SX, solid content (43 ± 1)%. Some performance parameters are shown in Tables 1 and 2.

Table 1. Performance parameters of SH.

Epoxy Resin EP-20 (%)	Polyethylene Glycol (%)	Epoxy Value	Proportion	Viscosity (mPa·s/25 °C)	Solid Content (%)	PH
25	25	0.20	1.10	≤2000	50 ± 1	7

Table 2. Performance parameters of SR.

Styrene (%)	Butyl Acrylate (%)	Methyl Methacrylate (%)	Methacrylate (%)	Viscosity (mPa·s/25 °C)	Solid Content (%)	PH
21.9	23.8	1.96	0.96	≤1800	45 ± 1	8

Functional filler: chopped CF, performance parameters are shown in Table 3. Conductive CB, performance parameters are shown in Table 4.

Table 3. Performance parameters of CF.

Diameter (μm)	Carbon Content (%)	Rate of Elongation (%)	Tensile Modulus (GPa)	Tensile Strength (MPa)	Electrical Resistivity (Ω·cm)	Density (10^3 kg/m^3)
≤8	97	2.1	230	4000	1.3	1.76

Table 4. Performance parameters of CB.

Product Feature	Specific Surface Area (m^2/g)	Electrical Resistivity (Ω·cm)	Particle Size (nm)	Density (kg/m^3)	PH
Black solid powder	230	101	15	144	8

2.2. Mix Proportion and Specimen Making

The existing research results of the research group show that the percolation threshold of 5 mm chopped CF is about 1.0% [32], and the percolation threshold of CB is about 2.0%. Therefore, in this paper, the CF blending ratio was 1.0%, and the CF was replaced at a 50% substitution rate, which ensured that the cost was reduced and the working performance was improved under the condition of constant conductivity. The content of the silica fume was 10%, which allowed for CF and CB to be more evenly dispersed in the cement matrix. Four kinds of polymers, SG, SR, SH, and SX, were selected to prepare polymer-modified cement paste. The P/C were 0% (S0 control group), 5%, 10%, 15%, and 20%, respectively. The water cement ratio was 0.55, the dispersant was 0.4%, the defoamer was 0.15%, the water reducing agent was 1.2%, and the expansion agent was 5%. The test mix ratio is shown in Table 5.

Table 5. Test mix ratio.

Group Number	CF (%)	CF (g)	CB (%)	CB (g)	P/C (%)	Polymer Emulsion (g)	Cement (g)	Water (g)
S0	0.5	6.5	1	13	0	0	1300	770
SG1	0.5	6.5	1	13	5	135	1300	716
SG2	0.5	6.5	1	13	10	271	1300	646
SG3	0.5	6.5	1	13	15	406	1300	575
SG4	0.5	6.5	1	13	20	542	1300	505
SR1	0.5	6.5	1	13	5	144	1300	707
SR2	0.5	6.5	1	13	10	289	1300	628
SR3	0.5	6.5	1	13	15	433	1300	548
SR4	0.5	6.5	1	13	20	578	1300	469
SH1	0.5	6.5	1	13	5	130	1300	722
SH2	0.5	6.5	1	13	10	260	1300	657
SH3	0.5	6.5	1	13	15	390	1300	592
SH4	0.5	6.5	1	13	20	520	1300	527
SX1	0.5	6.5	1	13	5	151	1300	700
SX2	0.5	6.5	1	13	10	302	1300	614
SX3	0.5	6.5	1	13	15	453	1300	528
SX4	0.5	6.5	1	13	20	605	1300	442

In this experiment, the ultrasonic dispersion method was used to improve the stirring process to prepare CF dispersion, and quantitative water was added to the constant temperature water bath to 60 °C. The 0.4% dispersant methyl cellulose was slowly added to water and stirred at a constant speed to fully dissolve it. Then, a calculated amount of conductive CF was added and fully stirred. During the stirring process, 0.15% defoamer was added to eliminate excess bubbles. Then, the prepared conductive CF dispersion was transferred to an ultrasonic cleaner and ultrasonically dispersed for 1 h to obtain a well-dispersed conductive CF dispersion. Then, according to DL/T 5126-2021 [33], cement, sand, CB, silica fume, expansion agent, etc., were dry-mixed in a cement paste mixer for 1 min; following this, the abovementioned CF dispersion was mixed with water reducing agent and water into the dry material, and then the polymer emulsion was added to the slurry. Mixing, stirring slowly for 120 s, pausing for 15 s, and then stirring quickly for 120 s to was performed to obtain a freshly mixed composite slurry. The specimens were formed at 40 mm × 40 mm × 160 mm dimensions, with copper mesh electrodes of 30 mm × 50 mm being embedded at 10 mm from both ends of the specimen and 40 mm from the center. All electrodes are inserted directly into the bottom of the specimen. The specimens were cured by dry–wet cycle method. After curing in the wet curing box for 24 h (+24 h), the specimens were demoulded. After 2 days of curing in the wet curing box, the specimens were placed in water at 20 ± 3 °C for 5 d, and then cured in the dry curing box for 21 d. The specimen making process is shown in Figure 1.

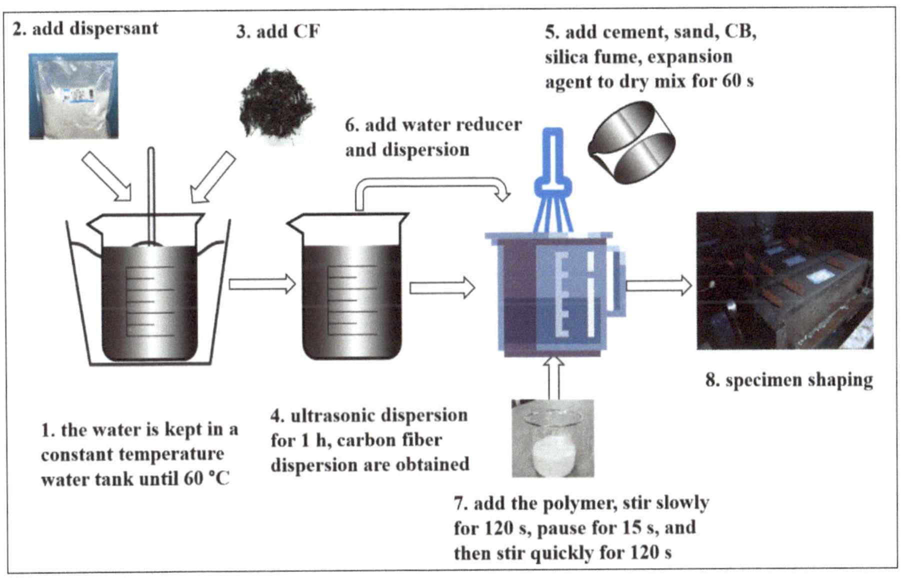

Figure 1. The specimen making process.

2.3. Test Method

The fluidity test was carried out according to GB/T 8077-2012 [34]. The water absorption test was carried out with reference to DL/T 5126-2021 [33]. The calculation formula of water absorption is shown in Equation (1).

$$W_A = \frac{G_1 - G_0}{G_0} \times 100\% \quad (1)$$

where W_A is water absorption (%), G_1 is the mass of the specimen after water absorption (g), and G_0 is the drying quality of the specimen (g).

The flexural strength and compressive strength test refer to DL/T 5126-2021 [33]. The resistivity was tested by the four-electrode method, and the schematic diagram is shown in Figure 2. The relationship between resistance value R, voltage value U, and current value I is shown in Equation (2). The calculation formula of the volume resistivity value (Ω·m) of the specimen is shown in Equation (3).

$$R = U/I \tag{2}$$

$$\rho = RS/L \tag{3}$$

where ρ represents the resistivity (Ω·m), R represents the resistance value (Ω), S represents the cross-sectional area of the specimen (m^2), and L represents the length between the outer electrodes in the direction of the specimen (m).

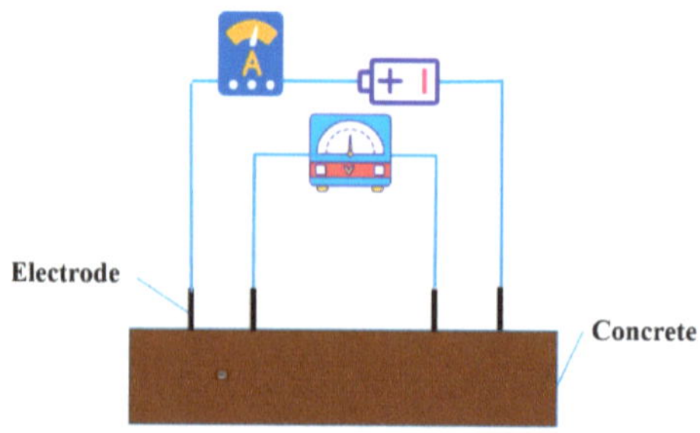

Figure 2. Resistivity test method (four electrode method).

3. Results and Discussion

3.1. Effect of Polymer on Physical Properties

The cement paste of the S0 control group without polymer was plastic, easy to pour, good in construction workability, and shows 24 h of stripping. Compared with the S0 group, the construction workability of the four cement pastes with polymer emulsion was the best in the SH group, followed by the SR group, and the membrane was removed after 48 h with high integrity. The construction workability of cement paste mixed with SX was the worst, which was dry and hard and could not meet the requirements of the specification. The low dispersibility of CF and the low hydrophilicity of CB, as well as the resulting huge polar interaction, led to the extreme aggregation of particles, which was the main reason for the poor fluidity and workability of the cement paste. However, the addition of SH and SR could significantly improve the delaying effect of conductive cement-based materials and maintain good construction workability due to good toughness, bonding strength, and water retention.

The effect of polymer on the fluidity of cement paste is shown in Figure 3. The fluidity of cement paste in the SH group showed an upward trend and was higher than that in the control group. The fluidity of cement paste in the SG, SR, and SX groups showed a downward trend and was lower than that in the control group. When the P/C is 15%, the fluidity of cement paste in the SH group and SR groups was 183 mm and 169 mm, respectively, which is 1.1% and 6.6% higher than that of the control group, indicating that SH and SR had little effect on the fluidity of cement paste. This is because the addition of polymer emulsion can introduce a large amount of gas, so that the material has better thixotropy and shear thinning properties, thereby improving the fluidity of cement paste. The addition of SH and SR can improve the negative effects caused by CF and CB, so that the conductive cement-based materials maintain good fluidity. Therefore, based on the influence of polymer on physical properties such as fluidity, setting time and workability, SH and SR are preferred.

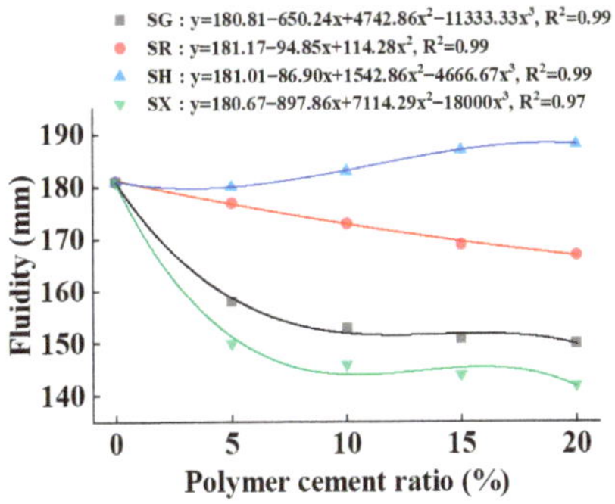

Figure 3. Fluidity of cement paste with different P/C.

3.2. Effect of Polymer on Mechanical Properties

The flexural strength to compressive strength (F/C) was selected to investigate the toughness of the specimen, so as to explore the influence of polymer on the toughening performance of cement paste. The effect of polymer on the F/C is shown in Figure 4.

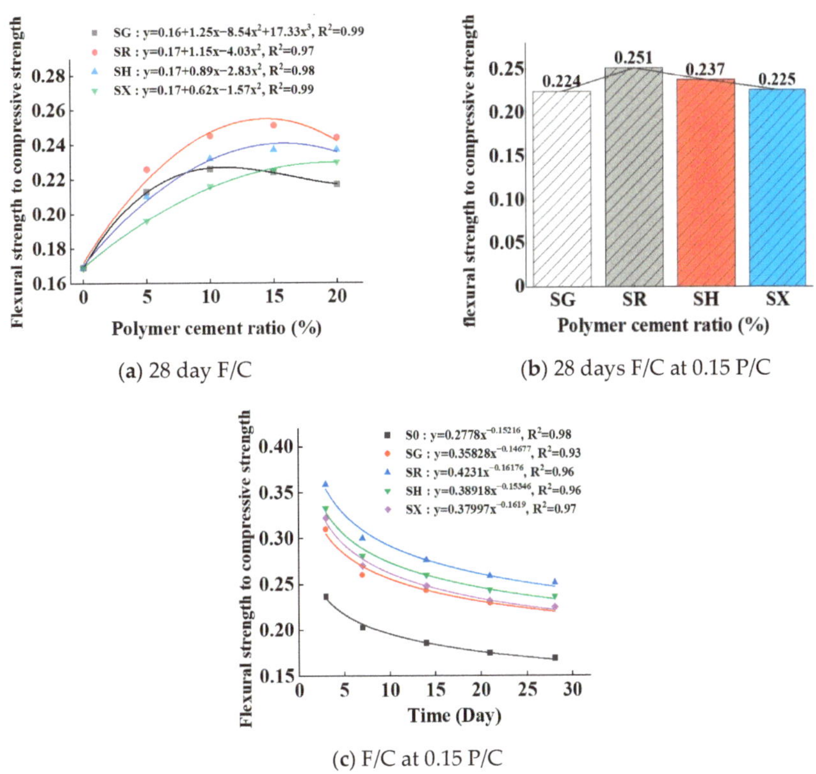

Figure 4. F/C with different P/C.

It can be seen from Figure 4a,b that the F/C of the four polymer test groups increased significantly with the increase in the P/C and then decreased slowly. Among them, the SR group was the best, followed by the SH group. When the P/C was 0.15, the peak value was reached. The folding pressure of the SR group was 48.5% higher than that of the control group, and the pressure of the SH group was 40.8% higher than that of the control group. The improvement in the mechanical strength of cement paste is due to the excellent flexibility and viscoelasticity of polymer chains, which can improve the brittle structure of cement-based materials. The polymer emulsion can increase the contact angle of the cement paste surface, effectively improve the flexural strength and tensile strength of the material, enhance the toughness, and improve the cracking phenomenon of the material.

It can be seen from Figure 4c that the F/C of the four polymer test groups decreased with the increase in age. The reason is that cement slurry dehydrated slowly after adding polymer. In the early stage of cement hydration, more dispersed polymer particles were adsorbed around the cement particles, significantly hindering the hydration of cement. Therefore, in the early stage of curing, the slurry exhibited slow setting, high flexibility, and a high F/C. With continuous hydration, the degree of hardening of the slurry increased, the flexibility decreased, and the F/C decreased. This is also the reason why polymer mortar needs to be brought into the curing box with a mold and needs special curing processes such as partial age water curing. It can be seen that the F/C of the four polymer test groups in the whole age are much higher than those of the control group, which fully shows the toughening advantages of the polymer on the cement-based materials, especially the SR group and the SH group. Therefore, based on the influence of polymer on F/C and toughness, SR and SH are preferred.

3.3. *Effect of Polymer on the Conductivity*

The effect of polymers on the conductivity of cement paste is shown in Figure 5. It can be seen from Figure 5a that the resistivity of cement paste in the SG, SR and SH groups decreased with the increase in P/C and gradually stabilized, indicating that the incorporation of polymer can improve the conductivity of cement paste. Among them, the SH group had the best conductivity, and the resistivity decreased by 43.0% compared with the control group at the P/C of 0.20, followed by the SR group, and the resistivity decreased by 20.3% compared with the control group at the P/C of 0.20. The reason is that after the polymer was added to the slurry, the matrix was denser, the cohesion between the cement hydration components was enhanced, the pores were reduced, the conductive network was more closely lapped, and the resistivity was reduced. In addition, the resistivity of the SX group increased slightly after reaching the lowest value at the P/C of 0.15, the resistivity of the SG group did not decrease, and the conductivity of the two groups was poor. It can be seen from Figure 5b that, except the resistivity of the SX group, which exceeds that of the control group, the resistivity of the other three groups decreases; the resistivity of the SH group decreases greatly, which obtained the optimal conductivity, followed by the SR group.

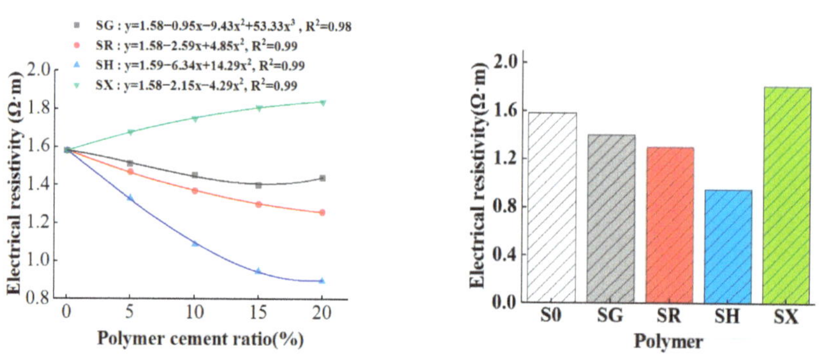

(**a**) Change in resistivity with P/C (**b**) 28 days electrical resistivity at 0.15 P/C

Figure 5. Resistivity of polymer cement paste.

The change in the resistivity of the specimen with age is shown in Figure 6. The resistivity of each polymer group and the control group increased with the increase in age and then gradually stabilized. The reason is that due to the high water content of the specimen in the initial stage of cement hydration, the material was in a liquid phase and solid phase blending environment, the conductivity was strong, and the resistivity was low. With the continuous hydration and hardening, the water content decreased, and the conductive network was completely constructed by solid phase conductive filler and cement hydration products. The resistivity rose and tended to be stable. Therefore, based on the influence of polymer on the conductivity, SH and SR are preferred.

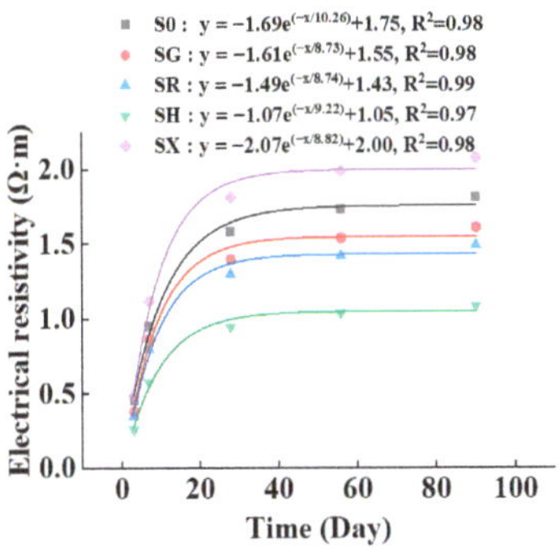

Figure 6. Change of resistivity with age at 0.15 P/C.

3.4. Effect of Polymer on Waterproof Performance

The effect of polymers on the waterproof performance of cement paste is shown in Figure 7. The water absorption rate of the four polymer groups decreased with the increase in the P/C, and the water absorption rate was SH < SR < SG < SX. The SH group and the SR group reached the lowest value when the P/C was 0.15, and the water absorption rate was reduced by 61.8% and 51.5%, respectively, compared with the control group. After that, the water absorption rate increased slightly with the increase in the P/C, indicating that the P/C of 0.15 was the best blending amount and that the optimal water absorption rate could be obtained. The reason is that the flocculation product generated by the emulsion particles of the polymer emulsion encapsulated the cement hydration product and filled the gap among the particles, which greatly improved the compactness of the cement slurry. When the emulsion particles were solidified into a film, a dense waterproof cementing layer was formed inside the slurry, so that the water absorption rate was reduced. When the P/C was 0.15, the water absorption rate reached the lowest value. However, with the further increase in the P/C, the water absorption rate increased. The reason is that the air content of the modified cement paste increased, so that the moisture absorption rate increased slightly. Therefore, based on the influence of polymer on water absorption, SH and SR are preferred.

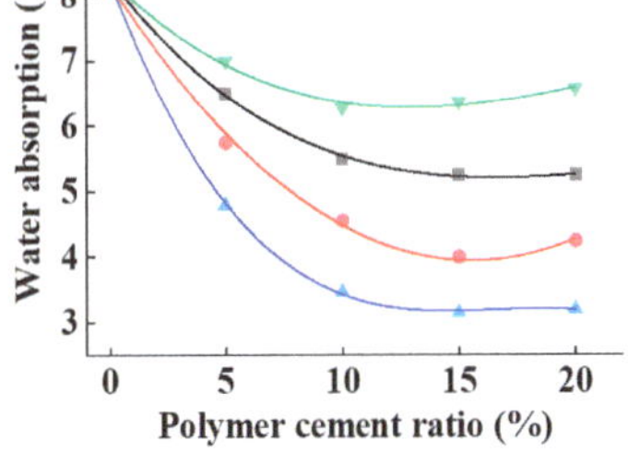

Figure 7. Water absorption of cement paste with different P/C.

3.5. Microscopic Analysis

The influence of polymer lotion on the microstructure of the conductive cement-based composite matrix is very critical. The SEM scanning microscopic morphology of the SH specimen is shown in Figure 8. Figure 8a shows the cement sample mixed with 5% SH and Figure 8b shows the cement sample mixed with 15% SH. It can be seen that the matrix of 15% polymer cement sample was denser, and there was more distinct membrane substance adhesion between cement hydration products, which made the matrix integrity stronger. The morphology of the 5% polymer sample was relatively dispersed, while the relatively dense matrix of the 15% polymer lotion sample made the material more waterproof, and also provided a good complete conductive matrix for electronic transition.

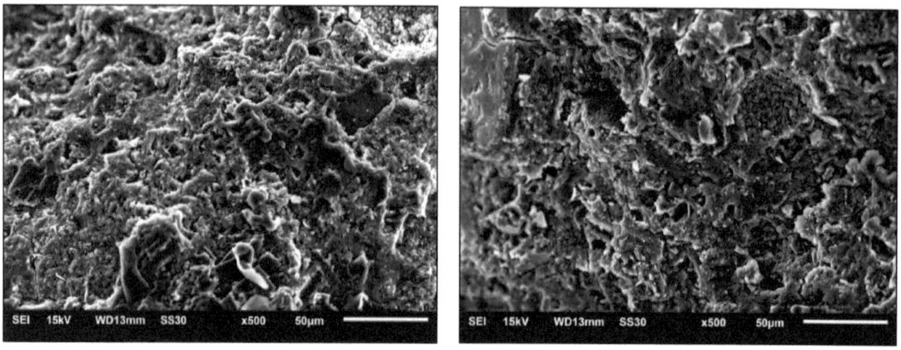

(**a**) SH with 0.05 P/C (**b**) SH with 0.15 P/C

Figure 8. SEM scanning diagram of SH specimen.

4. Conclusions

In this study, four polymer emulsions of SG, SR, SH, and SX and two conductive functional materials of CF and CB were used to prepare polymer conductive cement-based materials according to a certain proportion. The effects of different polymer–cement ratios of four polymers on the physical, mechanical, and conductive properties of conductive cement-based materials were studied experimentally. An SEM analysis of the specimens was carried out to explore the mechanism of polymer-modified conductive cement-based materials, and the modification effects of the four polymer emulsions were compared. According to the test results, the following conclusions are drawn.

(1) The fluidity of cement paste was improved by SH and increased with the increase in P/C. SG, SR, and SX all reduced the fluidity of the cement paste, and the reduced rate of SR was the smallest. All four polymers have a delaying effect, resulting in the hardening and peeling time of the specimens extending to 48 h.
(2) With the increase in flexural strength and the loss of compressive strength, the F/C of waterborne epoxy and SR-modified cement paste was increased to varying degrees, and the toughness of the material was improved. Both of them reached their maximum value when the P/C was 0.15. The improvement effect of SR was greater than that of SH.
(3) Except for SX, which can lead to cement paste resistivity increasing, the rest of the three kinds of polymer emulsion can reduce the resistivity of cement paste, which is beneficial to improving its electrical conductivity, with the improved effect of SH > SR > SG. Among them, the SH group and the SR group obtained the lowest resistivity at the P/C of 0.15.
(4) Four kinds of polymer emulsions can significantly reduce the water absorption of the specimens and improve the waterproof performance. The improvement effect is SH > SR > SG > SX. Among them, both the SH group and the SG group obtained the minimum water absorption at the P/C of 0.15.

(5) Microscopic morphology shows that a polymer-modified cement-based sample with a P/C of 0.15 can obtain a denser matrix, making the material more waterproof and providing a good and complete conductive network for electronic transitions.
(6) SH and SR selected in this experiment can be used to improve the toughness and waterproof performance of conductive cement-based materials, which is conducive to the preparation of cement mortar with high waterproof-performance requirements and low strength requirements. One can effectively use the electrothermal performance of conductive cement-based materials for electrothermal dehumidification, while improving the durability and corrosion resistance of materials. On this basis, our research group plans to further study the electrothermal dehumidification performance and corrosion resistance of polymer-modified conductive cement-based materials.

Author Contributions: Conceptualization, M.L.; methodology, J.Z.; software, G.L.; validation, F.C.; formal analysis, F.C.; investigation, J.Z. and Q.Z.; resources, Q.Z.; data curation, G.L.; writing—original draft preparation, M.L.; writing—review and editing, P.G.; visualization, Q.Z.; supervision, P.G.; project administration, M.L.; funding acquisition, P.G. All authors have read and agreed to the published version of the manuscript.

Funding: This research was funded by National Natural Science Foundation of China (grant number 52308470), Natural Science Basic Research Program of Shanxi (grant number 2023-JC-YB-375).

Data Availability Statement: The data presented in this study are available in this article.

Conflicts of Interest: The authors declare no conflict of interest.

References

1. Wang, R.; Gao, P.W.; Li, W.X.; Tian, Y.L.; Wang, L.M. Experimental study on dehumidification and chlorine-removal efficiency of conductive polymer mortar. *Constr. Build. Mater.* **2020**, *240*, 117950. [CrossRef]
2. Cui, S.P.; Liu, Y.X.; Lan, M.Z.; Wang, Z.M.; Wang, Y.L. *Research Status and Development Prospect of Conductive Concrete*; School of Materials, Beijing University of Technology: Beijing, China, 2005.
3. Gao, Y.X. *Preparation and Electrical Conductivity of Carbon Fiber Graphite Conductive Asphalt Concrete*; Wuhan University of Technology: Wuhan, China, 2011.
4. Sun, Y.F.; Gao, P.W.; Geng, F.; Li, H.R.; Zhang, L.F.; Liu, H.W. Thermal conductivity and mechanical properties of porous concrete materials. *Mater. Lett.* **2017**, *209*, 349–352. [CrossRef]
5. Huang, Y.H.; Rao, R.; Liu, C.H.; Liu, A.R. A Graphene Conductive Concrete. CN Patent 201610389015, 9 November 2016.
6. Lu, H. Summary of conductive concrete research. In *Fujian Quality Management*; Guangzhou University: Guangzhou, China, 2019; Volume 10, p. 152.
7. Lee, H.; Park, S.; Kim, D.; Chung, W. Heating Performance of Cementitious Composites with Carbon-Based Nanomaterials. *Crystals* **2022**, *12*, 716. [CrossRef]
8. Lee, H.; Seong, J.; Chung, W. Effect of curing time on thermal response characterization of carbon-nano cementitious composites. *Compos. Struct.* **2021**, *265*, 113781. [CrossRef]
9. Lu, Z.; Wu, X.; Yang, K. Study on the electrical conductivity of carbon black carbon fiber composite conductive concrete. *Concrete* **2014**, *11*, 81–83, 86.
10. Guo, C.H.; Tang, W.; Liu, S.H. Performance and microstructure of graphite powder conductive concrete. *Bull. Chin. Silic. Soc.* **2017**, *36*, 3174–3179.
11. Wang, Y.Y. Application of Steel Fiber Reinforced Concrete in Xiaoqi Hydropower Station. *Water Conserv. Heaven. Earth* **2013**, *7*, 38–40.
12. Yu, W.B.; Li, S.Y.; Feng, W.J.; Yi, X. Analysis of the current situation and development trend of road snow melting and deicing technology. *Permafr. Glacier* **2011**, *33*, 933–940.
13. Zhang, J. Research status of conductive concrete for snow melting and deicing of highway bridges. *Shanxi Archit.* **2007**, *33*, 267–268.
14. Mu, C.P. Application of polymer modified cement composites in high-rise building materials. *Chin. Dwell.* **2014**, *3*, 183.
15. Zhou, R.; Wang, J.R.; Zhang, H.B. Research progress of polymer modified cement-based materials. *New Chem. Mater.* **2021**, *49*, 275–279.
16. Hou, Z.F.; Li, Z.Q.; Tang, Z.Q. Study on the electrical conductivity of carbon fiber reinforced concrete for snow melting and ice melting. *J. Wuhan Univ. Technol.* **2002**, *24*, 32–34, 66.
17. Liu, S.; Ge, Y.; Wu, M.; Xiao, H.; Kong, Y. Properties and road engineering application of carbon fiber modifie-electrically conductive concrete. *Struct. Concr.* **2021**, *22*, 410–421. [CrossRef]

18. You, I.; Lee, S.J.; Zi, G.; Lim, D. Influence of Carbon Fiber Incorporation on Electrical Conductivity of Cement Composites. *Appl. Sci.* **2020**, *10*, 8993. [CrossRef]
19. Ding, T.; Liu, Y. Study on the Produce Process and Electrical Resistivity of Carbon Fiber Conductive Concrete. *Adv. Mater. Res.* **2014**, *960–961*, 14–17. [CrossRef]
20. Gull, I.; Tantray, M.A. Dispersion of electrically conductive carbon fibres in self-compacting concrete using chemical and mechanical dispersing techniques. *Int. J. Microstruct. Mater. Prop.* **2020**, *15*, 242–266. [CrossRef]
21. Dehghanpour, H.; Yilmaz, K.; Ipek, M. Evaluation of recycled nano carbon black and waste erosion wires in electrically conductive concretes. *Constr. Build. Mater.* **2019**, *221*, 109–121. [CrossRef]
22. Dong, W.; Li, W.; Shen, L.; Sheng, D. Piezoresistive behaviours of carbon black cement-based sensors with layer-distributed conductive rubber fibres. *Mater. Des.* **2019**, *182*, 108012. [CrossRef]
23. Ohama, Y. Polymer-based admixtures. *Cem. Concr. Compos.* **1998**, *20*, 189–212. [CrossRef]
24. Long, J.; Yu, K.; Li, G.D. Interaction between polymer and cement hydrate. *Concrete* **1995**, *3*, 35–37, 41.
25. Schulze, J. Influence of water-cement ratio and cement content on the properties of polymer-modified mortars. *Cem. Concr. Res.* **1999**, *29*, 909–915. [CrossRef]
26. Wang, P.M.; Xu, Q.; Stark, J. Mechanical properties of styrene-butadiene latex modified cement mortar for bridge deck. *J. Build. Mater.* **2001**, *1*, 1–6.
27. Bureau, L.; Alliche, A.; Pilvin, P.; Pascal, S. Mechanical characterization of a styrene–butadiene modified mortar. *Mater. Sci. Eng. A Struct. Mater. Prop. Microstruct. Process.* **2001**, *308*, 233–240. [CrossRef]
28. Pascal, S.; Alliche, A.; Pilvin, P. Mechanical behaviour of polymer modified mortars. *Mater. Sci. Eng. A Struct. Mater. Prop. Microstruct. Process.* **2004**, *380*, 1–8. [CrossRef]
29. Yang, X.X. *Basic Research on the Application of Polymer Modified Cement Mortar*; Fuzhou University: Fuzhou, China, 2006.
30. Aggarwal, L.K.; Thapliyal, P.C.; Karade, S.R. Properties of polymer-modified mortars using epoxy and acrylic emulsions. *Constr. Build. Mater.* **2007**, *21*, 379–383. [CrossRef]
31. Tian, Y.L.; Gao, P.W.; Wang, R.; Wang, L.M.; Zhong, J.J.; Li, J.S. Study on the influence of organic polymers on the physical and mechanical properties of cement-based materials. *IOP Conf. Ser. Earth Environ. Sci.* **2020**, *474*, 072028. [CrossRef]
32. Chen, M.; Gao, P.W.; Geng, F.; Zhang, L.F.; Liu, H.W. Mechanical and smart properties of carbon fiber and graphite conductive concrete for internal damage monitoring of structure. *Constr. Build. Mater.* **2017**, *142*, 320–327. [CrossRef]
33. *DL/T 5126-2021*; Polymer Modified Cement Mortar Test Procedure. National Energy Administration: Beijing, China, 2021.
34. *GB/T8077-2012*; Test Method for Homogeneity of Concrete Admixture. State Administration of Quality Supervision, Inspection and Quarantine of the People's Republic of China: Beijing, China; China National Standardization Management Committee: Beijing, China, 2012.

buildings

MDPI

Article

Molecular Dynamics Study on Interfacial Strengthening Mechanisms of Ettringite/Polymer Nanocomposites

Liwei Zhang, Heping Zheng * and Huilin Xie

Department of Civil Engineering, Qingdao University of Technology, Qingdao 266033, China; zhangliwei@qut.edu.cn (L.Z.); 17863527946@163.com (H.X.)
* Correspondence: zhengheping0802@126.com

Abstract: Compared with polymer-modified ordinary-Portland-cement-based materials, research on cement materials based on polymer-modified sulfoaluminate is still in the preliminary stage and lacks an understanding of the mechanism of the interaction interface. The aim of this work is to study the bond performance of ettringite, the main hydration product of sulfoaluminate cement, with various types of polymers using molecular dynamics methods. Steered molecular dynamics were used to simulate the separation of polyamide (PA), polyethylene glycol (PEG), polyacrylic acid (PAA) and polypropylene (PP) from ettringite substrate, reflecting the order of bond properties of the four polymers: PAA > PA > PEG > PP. The internal mechanism of bond properties between different polymers and ettringite was analyzed by studying the local structure and dynamic characteristics. The results show that a Ca–O ionic pair is formed between the calcium ions on the surface of the polymer and ettringite substrate, resulting in strong interaction. In addition, the formation of a H bond also contributes to bond performance. The properties of the polymer itself, such as the degree of polymerization and branched-chain freedom, affect the coordination of the polymer to the substrate. This study provides valuable insights for advancing the development of polymer-modified sulfoaluminate-cement-based materials.

Keywords: polymer; ettringite; interfacial bonding performance; molecular dynamics

Citation: Zhang, L.; Zheng, H.; Xie, H. Molecular Dynamics Study on Interfacial Strengthening Mechanisms of Ettringite/Polymer Nanocomposites. *Buildings* **2023**, *13*, 2976. https://doi.org/10.3390/buildings13122976

Academic Editor: Rajai Zuheir Al-Rousan

Received: 3 November 2023
Revised: 24 November 2023
Accepted: 27 November 2023
Published: 29 November 2023

1. Introduction

Portland cement has been widely used because of its reliable performance and low price. However, it also has problems, such as long maintenance time, slow development of early strength, poor fluidity and low compaction rate. Therefore, sulfoaluminate cement, which has the characteristics of fast hardening, early strength, shrinkage and short design age, has become a research focus of new building materials in recent years and is widely used in construction, roads, water conservancy, port tunnels and other fields of emergency repair. Compared with Portland cement, the CO_2 content of sulfoaluminate cement can be reduced by 40% per ton [1], which is in line with the development trend regarding low carbon emissions and high-performance cement. However, as a brittle material, cement has poor bending and tensile effects, and also has certain defects in wear resistance and flexibility, which cannot meet the requirements for its toughness and bond strength. Fortunately, polymer-reinforced cement matrix composites can solve this problem well.

Polymer materials exhibit excellent toughness and high bond strength, attributed to the relaxation movement between their chain segments. The incorporation of polymers into cement-based materials results in the formation of a polymer–cement composite system. On one hand, polymers can adsorb particles generated during cement hydration through their inherent polar functional groups, creating a network with interpenetrating structures of polymer–cement hydration products. On the other hand, the extended chain segments of polymers contribute to enhancing the stiffness of cement-based materials. This comprehensive enhancement significantly improves the toughness and bonding properties of cement-based materials. Because polymer-reinforced cement matrix composites are still cement-based, they are more compatible with concrete substrates. Polymer-reinforced cement matrix composites have been more widely used in projects such as the Yangtze River Three Gorges Power Station, the National Theatre and municipal projects in China [2,3].

Currently, the research on polymer-modified sulfoaluminate cement is still in the preliminary stage. Scholars have carried out research on the mechanical properties, workability and toughness of polyacrylic acid and polyvinyl alcohol for sulfoaluminate-cement-based materials. The experiments have demonstrated that the incorporation of polymers effectively improves the mechanical properties of cement-based materials, and compounding could play a synergistic role to improve the flowability, increase flexural and compressive strength and enhance toughness [4–6]. Xiao et al. [7] found that the optimal dosage of polyvinyl alcohol was 0.75%, and the 28d flexural strength of mortar increased by 15% under this dosage. Zhang et al. [8] studied the effects of polyacrylamide on the setting time, mechanical properties and water resistance. The study showed that the addition of polyacrylamide could significantly prolong the setting time and improve its fluidity, among which the 3d and 28d flexural strengths increased by 75% and 39%, respectively, when the dosing amount was 0–0.3%, and the mechanical properties were greatly improved. In addition, polymers such as polyacrylic acid, polyurethane and epoxy [9–11] can improve the performance of sulfoaluminate cement very well. According to previous studies, it can be found that the macroscopic properties, such as the mechanics of polymer-reinforced cement matrix composites, are mainly studied, while the study of the interfacial interaction between the two phases at the nanoscale is still imperfect and challenging, occurring only through experimental studies, where molecular dynamics has a significant advantage [12,13].

Molecular dynamics (MD) simulation can break the limitations of time and conditions, demonstrate experimental laws at the atomic level, realize the exploration of mechanisms and can also combine with the experimental development of novel materials [13,14]. Molecular dynamics has been used to elaborate and reveal the integrated interfacial mechanism between polymers and hydrated calcium silicate (C–S–H) [15]. Wang et al. [16] revealed the reasons for the differences in interfacial shear strength between different polymer fibers and C–S–H by the molecular dynamics method. Hou et al. [17] proved that the interfacial bonding between C–S–H and polymers is related to the polarity of polymer functional groups, H bonds, Ca–O ionic pairs and bond stability. Liu et al. [18] found that electrostatic and H-bond interactions increased the adhesion properties of cellulose fiber to the C–S–H interface. Zhou et al. [19,20] investigated the interaction mechanism of PVA with the micro-test method and MD simulations. However, there are significant differences in the composition of ordinary Portland cement and sulfoaluminate cement. The strength of ordinary-Portland-cement-based material is derived from the interaction between C–S–H gels, while the strength of sulfoaluminate-cement-based material is derived from ettringite crystals, which account for about 65% of the hydration products. Therefore, there may be differences in the affinity and bonding properties of the polymers to the two types of cement.

Due to the lack of targeted studies on the bonding performance of polymers with an ettringite interface, in this work, the bond performance of ettringite, the main hydration product of sulfoaluminate cement, with four types of polymers was investigated via the molecular dynamics method for the first time. The tensile strength of the four polymers separated from the ettringite substrate was simulated separately by tensile dynamics to reflect the bonding performance between the four polymers and ettringite. The intrinsic mechanism of bonding performance between different polymers and ettringite was analyzed by local structural and dynamic characterization. This work is instructive for the development of polymer-modified sulfoaluminate-cement-based materials.

2. Simulation Method

2.1. Model Construction

The initial model for the study is based on existing modeling of ettringite utilizing crystals from a crystal library [21]. Ettringite, with the molecular formula $3CaO{\cdot}Al_2O_3{\cdot}3CaSO_4{\cdot}32H_2O$ and a basic structural unit of $[Ca_3Al(OH)_6{\cdot}12H_2O]^{3+}$, exhibits a columnar structure depicted in Figure 1a [22–24]. These columns consist of lines of $[AI(OH)_6]^{3-}$ octahedra along the (001) direction, connected to neighboring columns by three groups of Ca^{2+} ions, completing coordination polyhedra with water molecules. Each Ca^{2+} ion is 8, coordinated by 4 H_2O molecules and 4 OH^- ions, forming a trigonal prism. The polyhedra share edges with adjacent $[AI(OH)_6]^{3-}$ octahedra. $SO_4{}^{2-}$ ions and remaining water molecules are hexagonally arranged in the cylindrical gap, encircling the positively charged cylinder's axis.

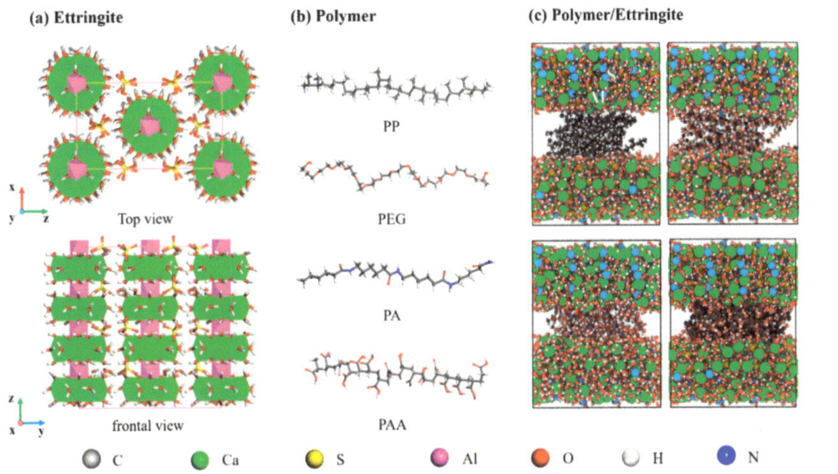

Figure 1. (**a**) Crystal structure model of ettringite; (**b**) four polymer models; (**c**) composite structure model.

Secondly, four representative polymers with different functional group types are selected, which are non-polar polypropylene as the control, polyethylene glycol with hydroxyl functional group, polyamide with amide functional group and polyacrylic acid with carboxyl functional group. Under the condition of the same number of functional groups, the molecular model is established by selecting the degree of polymerization of polypropylene polyethylene glycol and polyacrylic acid as 16. Due to the large molecular formula of polyamide monomer, on the premise that the main chain length of polyacrylic acid is basically the same as that of the other three polymers, the degree of polymerization is four. Their molecular structure formula is shown in Figure 1b.

The ettringite model of the bottom and top layers is created by means of the super cell, and the parameters are as follows: a = 44.67 Å, b = 38.68 Å, c = 25.87 Å. The polymer is left in 25 Å space between the two parts of ettringite. The four polymer molecules are placed in the middle of the ettringite in two layers of four for each layer to form the composite structural model of the sandwich. In addition, a vacuum layer of 80 Å is reserved at the top to prevent ettringite from crossing periodic boundaries and causing interaction between upper and lower layers, and the simulation box is large enough to accommodate the bonding and stripping behavior of ettringite and polymer. The balanced structures of the four materials are shown in Figure 1c.

2.2. Simulation Details

In this study, the ClayFF [25] force field is employed to characterize the interactions within ettringite substrates, offering a detailed representation of their structural dynamics.

Simultaneously, the CVFF force field is harnessed to delineate the interactions occurring within polymers [26]. The ClayFF has been proved to accurately describe the interaction of clay hydroxide and interface structure and has also been successfully applied in the molecular dynamics study of the interaction between cement-based materials. Ettringite is one of the hydration products of cement-based materials. The ClayFF force field does not provide for sulfate interactions, so sulfate parameters from Cannon [27,28] are used as supplements, with sulfur charges of 2e and oxygen charges of −1e. The non-bond interaction parameters and the charges of each atom in ettringite are referenced from Ref. [25]. The comprehensive simulation, executed using Lammps [29], incorporates the efficient descent method to systematically relax the entire system, aiming to minimize overall energy. Following system relaxation, an equilibrium adsorption phase is conducted under the NVT ensemble for a duration of 3 nanoseconds. To facilitate subsequent microstructure analysis, trajectory data are output at regular intervals of 1 picosecond (lps) throughout the simulation period. This meticulous approach ensures a thorough exploration of system dynamics, allowing for in-depth examination and interpretation of the resulting microstructural features.

At the stage of steered molecular dynamics, in the NVT system [30–32], an external force of the form shown below is applied to the centroid portion of the upper ettringite:

$$F = K((X_0 + V \times t) - X_{com})$$

where F is external force; K is the spring force constant, set as 0.0005 kcal/mol/nm^2; X_0 is the initial Z coordinate of the selected upper ettringite core [33]. V is the pulling speed of the reference point along the Z axis, set as 100 m/s. T is the simulation time; X_{com} is the selected dynamic location of the upper ettringite core along the Z direction. The schematic diagram of the polymer and ettringite matrix model under the action of additional forces is shown in Figure 2. To guarantee the precision of the tensile process simulation, a meticulous approach is adopted. The simulation step is finely tuned to 0.1 femtoseconds, and the simulation time is extended to 20 nanoseconds. This meticulous setup aims to capture the intricate details of the dynamic tensile behavior. Throughout the simulation process, data for force (F) and the center of mass displacement (X_{com}) are meticulously extracted. This extraction process enables the derivation of a comprehensive tensile force–displacement curve. By employing these finely tuned parameters and extracting key data points, the simulation aims to provide a detailed and accurate representation of the tensile behavior, facilitating a thorough analysis of the material's response under tensile loading conditions. In this study, Visual Molecular Dynamics (VMD) is utilized for graphical visualization [34].

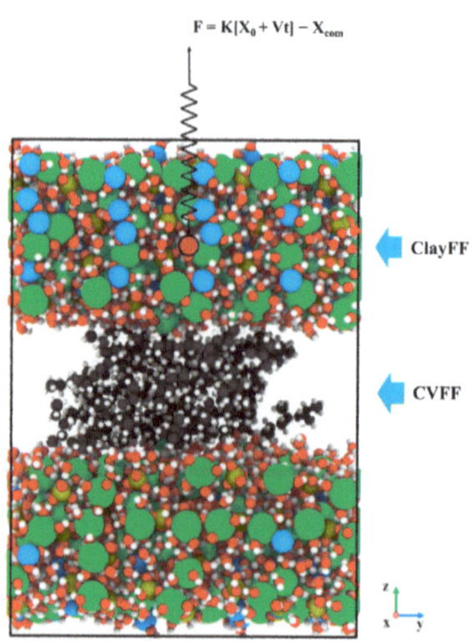

Figure 2. The steered molecular dynamics models of PA/ettringite.

3. Results and Discussion

The pulling force of the separation of the four polymers and ettringite substrate is simulated by steered molecular dynamics to reflect the interfacial bond property between the four polymers and ettringite. The microscopic mechanism of the difference in interfacial bonding properties is analyzed by local structural and dynamic characterization.

3.1. Separation Behavior

In order to preliminarily observe the behavior changes of four polymers, we extract the behavior change images of ettringite and four polymers during tensile dynamics. Simulation snapshots of the four systems are shown in Figure 3, which intuitively reflect the adsorption and separation behavior of the polymers.

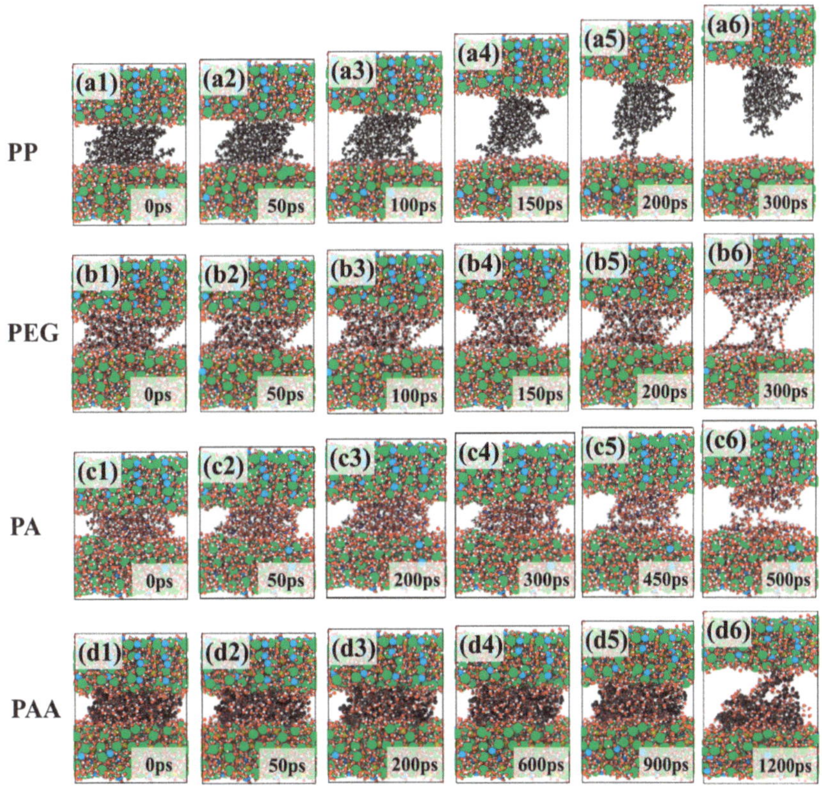

Figure 3. (**a1–a6**), (**b1–b6**), (**c1–c6**), and (**d1–d6**) indicate the damage snapshots of Ettringite/PP, Ettringite/PEG, Ettringite/PA, and Ettringite/PAA at different times, respectively.

The image comparison shows that the interface region between the four polymers and ettringite is separated with the increase in stretching time. PP is the first to detach, followed by PEG, PA and PAA. The bond strength relationship between the four polymers and ettringite can be qualitatively reflected from the simulated snapshots as PAA > PA > PEG > PP.

3.2. Force–Displacement Curves

Based on the qualitative analysis in the preceding section, the mechanical curve is extracted for quantitative analysis. Figure 4 shows the tensile force–displacement curves of the four polymers at a simulated tensile speed of 100 m/s using the tensile dynamics method. The slope of a curve is the ratio of the stretching force to the displacement at the stretching point. When the tensile force does not reach the critical value of the interfacial force, the value is positive. When the tensile force reaches the maximum value, that is, after the polymer is separated from the ettringite base, the slope becomes negative and remains unchanged.

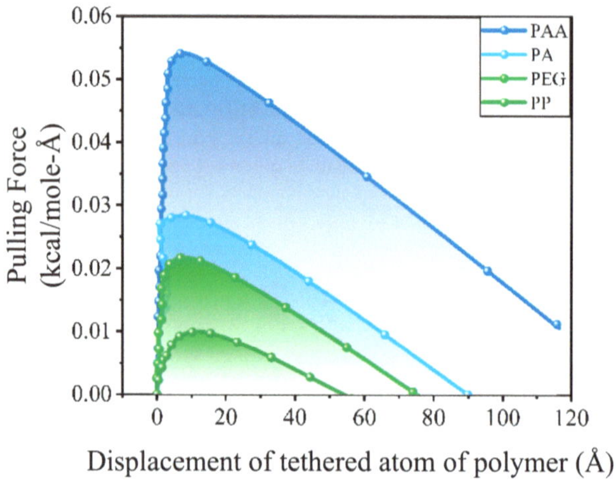

Figure 4. The ratio of the pulling force of the four polymers to the displacement.

It can be found from the curve of tensile dynamics simulation that the interface bond properties of four polymers and ettringite from high to low at the same tensile speed of different polymers are as follows: PAA > PA > PEG > PP. The first half of the curve is wavy because the chemical bond between the polymer and ettringite base is not completely disconnected in the tensile process, and the phenomenon of reconnection occurs.

The above laws are in good agreement with the experiment. Bezerra et al. [6,35] studied PVA and PP, both enols, and showed that the strength and toughness of PVA were higher than those of PP, and PVA had better adhesion to the substrate. At present, there are experimental studies on the properties of the four types of polymers modified, but there is a lack of horizontal comparison of the four polymers. Therefore, this work can effectively guide the selection of polymers and the development of experiments from the perspective of simulation. In the following sections, the mechanisms behind the differences in bonding performance between various polymers and ettringite were analyzed by the local structural and dynamic characteristics.

3.3. Atomic Intensity Distribution

The atomic intensity distribution can reflect atoms of the polymer on the surface of the ettringite. Figure 5a–d describe the intensity distribution of polyamide, polyacrylic acid, polyethylene glycol and the components between the four polymers, respectively, where Ca denotes the ettringite surface calcium atom, Os denotes the polymer single-bond oxygen atom, Od denotes the polymer double-bond oxygen atom, Oco denotes the polymer carboxyl oxygen atom, Oc denotes the polymer main-chain oxygen atom, Hal denotes the aluminum hydroxyl group of ettringite and C denotes the polymer main-chain carbon atom. While ensuring the same length of the main chains of the four polymers, the atomic content of the main chains of the polymers differed due to the degree of polymerization, so we mainly observe the changes in the intensity distribution of each polymer atom in terms of position.

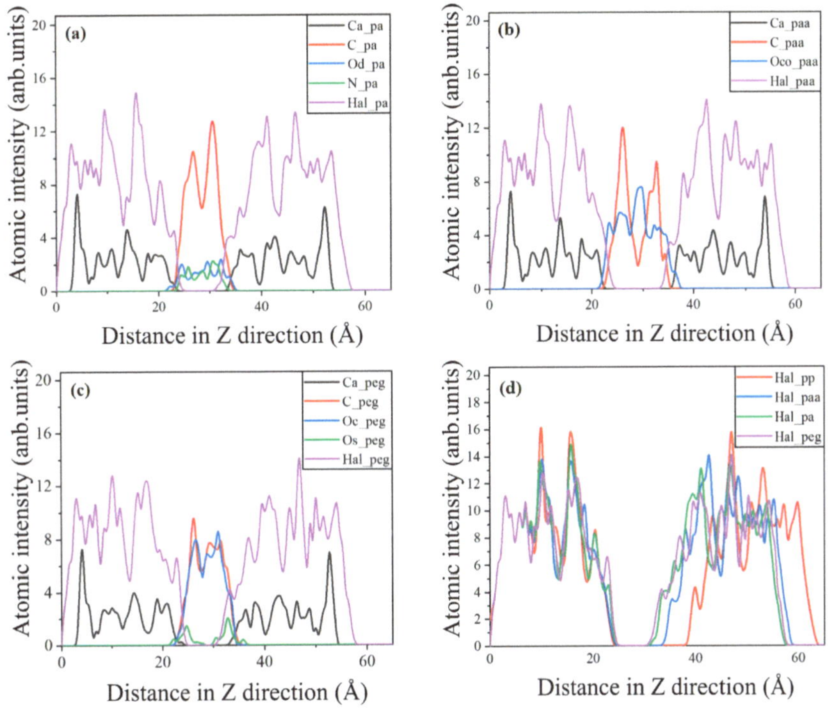

Figure 5. Atomic intensity distribution of different composite material systems: (**a**) Ettringite/PA, (**b**) Ettringite/PAA, (**c**) Ettringite/PEG, (**d**) Hydrogen atoms in four polymers.

It can be seen from Figure 5a that the Od intensity distribution is closer to the Ca layer than its main-chain carbon atoms at the ettringite interface, indicating the strong interaction between calcium atoms near the interface between polyamide and ettringite, which causes its distribution to shift. The atomic strength distribution of polyacrylic acid and ettringite substrate is shown in Figure 5b. The intensity peak of Oco of the carboxyl functional group of PAA is close to that of Ca, and, from the overall view of the polymer, the carboxyl Oco is biased to be distributed on the side of the main-chain C close to the substrate. Both phenomena can reflect the interaction of PAA with Ca of the surface layer of ettringite. Different from the former two kinds of polymers, polyethylene glycol is out of the side chain of hydroxyl functional groups, and the main chain also exists in the more electronegative Oc. From Figure 5c, the intensity distribution can be observed; the intensity peak of hydroxyl Os is very close to the intensity peak of Ca, both with strong interaction, while the main-chain oxygen, due to the restrictions of the main chain itself, does not show a bias towards the base. In addition to the interaction mentioned above, it can be obviously found from Figure 5d that Hal of the ettringite substrate of the PAA, PA and PEG system is more inclined to the intermediate polymer layer than the PP system, indicating that there is H-bond interaction between the three polymers and the substrate. The specific bonding situation will be analyzed later.

Fundamentally, due to the polarity of the polymer itself, the above three polymers interact with the surface calcium and hydroxyl groups of the ettringite substrate, reflecting the adsorption characteristics of the polymer and the substrate.

3.4. Local Interface Analysis

In order to accurately analyze the bond situation at the interface between polymer and ettringite, and exclude the influence brought by the complexity and diversity of the intermediate polymer layer, a single polymer at the interface was selected for subsequent analysis.

3.4.1. Interfacial Ca–O Ionic Pair

The radial distribution function (RDF) can qualitatively reflect the interface structure of polymers and ettringite [36]. Figure 6a–d show the bonding of the three polymer functional groups, PAA, PA and PEG, to the surface calcium of ettringite. In the PA/ettringite interface, the first peak of Ca–Od is 2.55 Å, which is consistent with the bond length of the calcium–oxygen polyhedron in ettringite crystal between 2.35 and 2.75 Å [37]. Therefore, the sharp peak of PA indicates that the double-bonded oxygen on the amide group is bonding with the surface Ca, and the peak value is high, indicating the strong interaction between the two. Figure 6e is a partial schematic diagram of the Ca–O ionic pair of PA and ettringite, which intuitively shows the bonding situation of the interface. This simulation phenomenon is consistent with the experimental characterization of the polymer in the process of cement modification and will participate in the chemical reaction regarding the formation of the Ca–O ionic pair [11,38,39].

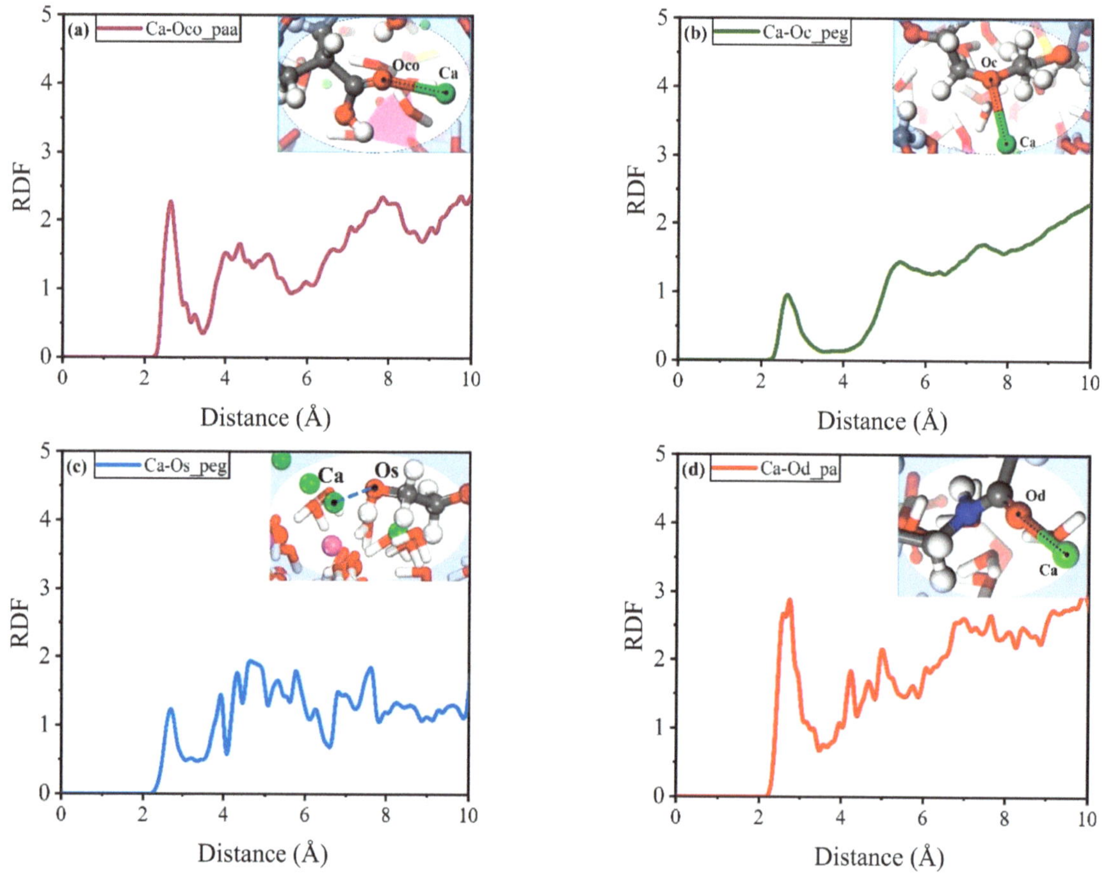

Figure 6. *Cont.*

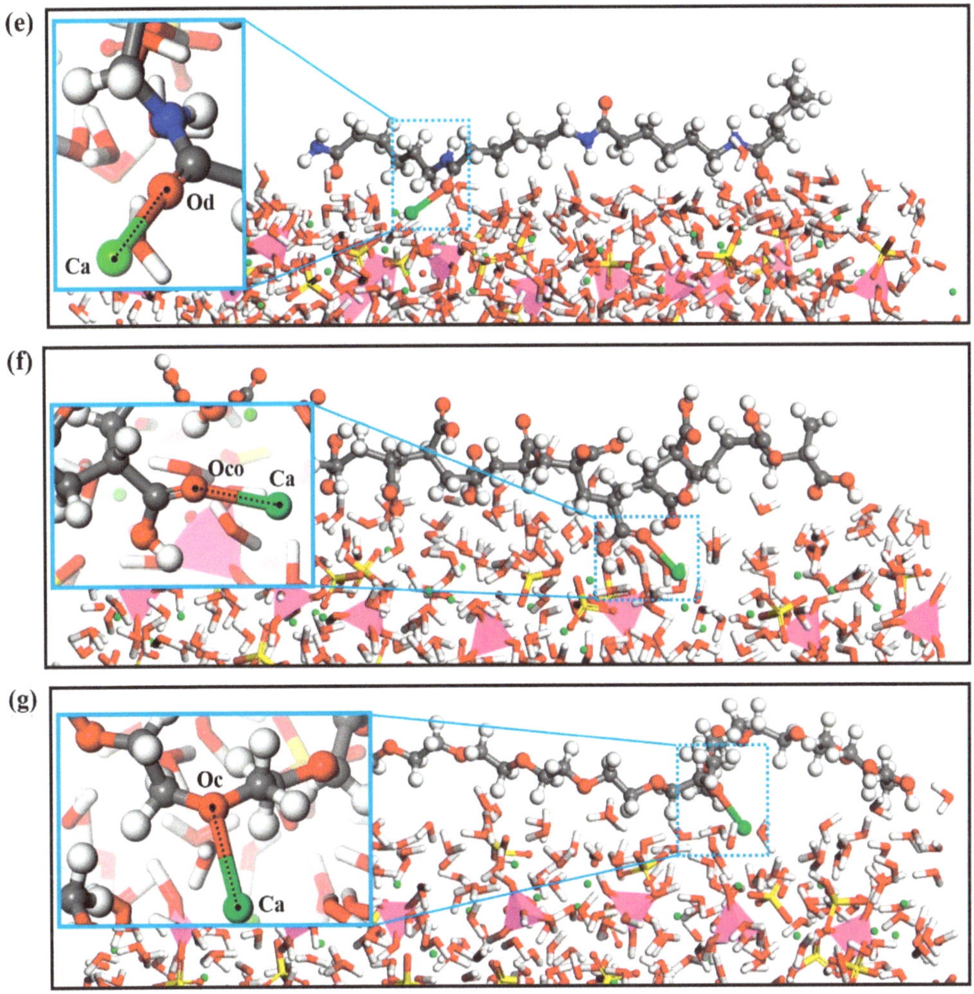

Figure 6. (**a–d**) The RDF of Ca–O; (**e**) local schematic diagram of Ca–O ionic pair at PA/ettringite interface; (**f**) local schematic diagram of Ca–O ionic pair at PAA/ettringite interface; (**g**) local schematic diagram of Ca–O ionic pair at PEG/ettringite interface. The blue box shows a partial enlargement of the chemical bond.

Similarly, the first peak of Ca on the surface of PAA and ettringite is about 2.65 Å, indicating that oxygen atoms of the carboxyl group in PAA can interact with calcium atoms on the ettringite interface. In PEG/ettringite composites, the first peaks of Ca–Os and Ca–Oc are both at 2.65 Å, respectively, indicating that the surface Ca of ettringite interacts not only with the hydroxyl group at the end of the PEG chain but also with the oxygen atom on the PEG main chain. This analysis method is consistent with the related simulation studies on polymers and CSH, indicating that polymer functional groups can interact with the substrate to achieve performance improvement [15,16,40,41]. The bonding between PAA, PEG and the ettringite interface is shown in Figure 6f,g.

The coordination number can be used to quantitatively analyze the interaction between different polymers and ettringite [36]. Table 1 shows the number of coordination calcium atoms of each polymer. The three polar polymers can coordinate with the surface calcium of ettringite, and the overall order of the total coordination number of the three polymers is PAA > PA > PEG. The coordination number of PAA is three times that of PA and PEG, indicating that PAA has the stronger binding ability with ettringite surface, mainly because the carboxyl molecular structure of the PAA branch chain is larger and the degree of freedom is higher, which makes it easier to capture the surface calcium of ettringite.

Table 1. The coordination number of calcium ions between each polymer and ettringite substrate.

CN	Ettringite/PA	Ettringite/PAA	Ettringite/PEG
Ca–Od	1.30	/	/
Ca–Os	/	/	0.58
Ca–Oco	/	4.38	/
Ca–Oc	/	/	0.66
Total	1.30	4.38	1.24

While ensuring that the main chain length of all kinds of polymers is basically the same, the degree of polymerization of PA is only one-fourth that of other polymers, and the number of functional groups is significantly smaller; thus, its value on the number of coordination calcium atoms is relatively small. In addition, the number of oxygen atoms in the main chain of PEG is much larger than that of hydroxyl oxygen at the end of the chain, but there is little difference between them in the coordination number, indicating that the main-chain oxygen atoms are difficult to bind to the calcium ions due to the limitation in the degree of freedom, and calcium prefers to adsorb on the hydroxyl oxygen. Thus, it can be seen that the polymerization degree, branched chain degrees of freedom and other properties of the polymer will affect the coordination of the polymer to the substrate.

3.4.2. H Bond

Apart from the crucial Ca–O ionic pair, the interaction between ettringite and the polymer involves H bonds that significantly influence the overall bond performance at the interface. In Figure 7, the Radial Distribution Function (RDF) diagram and local structure diagram elucidate the H-bond interactions between the polymer and ettringite. Here, H represents the hydrogen in the polymer functional group, Ow denotes water oxygen, Os signifies sulfate oxygen and Oal represents the oxygen in the aluminum hydroxyl group. The ettringite surface, comprising water molecules, aluminum hydroxyl groups and sulfate, facilitates the formation of two types of H bonds with the polymer. One type involves the oxygen atom as the acceptor, capable of accepting electrons from the hydrogen atom in the polymer functional group. The other type features the hydrogen atom as the donor, providing electrons to the oxygen atom of the polymer functional group. This dual role of H bonding underscores the intricate nature of the interactions at the polymer–ettringite interface, emphasizing the diverse mechanisms contributing to the overall bond performance. The detailed analysis presented in the RDF and local structure diagram enhances our understanding of the complex intermolecular dynamics governing the interface.

Figure 7a is the RDF curve of the PAA system. It can be seen from the figure that PAA can form five types of H bonds with the base within the bond formation range of 2.45 Å, and the H bonds formed are mainly with oxygen atoms of sulfate and oxygen atoms in water. The local structure diagrams of five kinds of bonds are shown in the blue box in Figure 7b. The peak position of the RDF diagram shows that only two representative polymers are indicated to show the H bond between the polymer and ettringite substrate more intuitively and briefly. Figure 7b shows the distribution of H bonds of all PAAs. It can be found that H bonds can also be formed among PAA, which increases the bond between polymers and increases the bond strength between PAA and the ettringite interface. In Figure 7c for the RDF curves of the PA system, PA is special in that the H-bond receptor comes from the amino nitrogen atoms, has less electronegativity than O atoms and, coupled with the number of functional groups, was only about a quarter of the PAA, decreasing the likelihood to form H bonds, so the hydrogen atoms form a H bond on the functional groups of the aft peak position, and the peak fell. The basket on the right of Figure 7d is the local diagram of H bonds. Similarly, there are H bonds between PA chains. It can be seen intuitively that the number of H bonds in PA decreases due to the small number of functional groups.

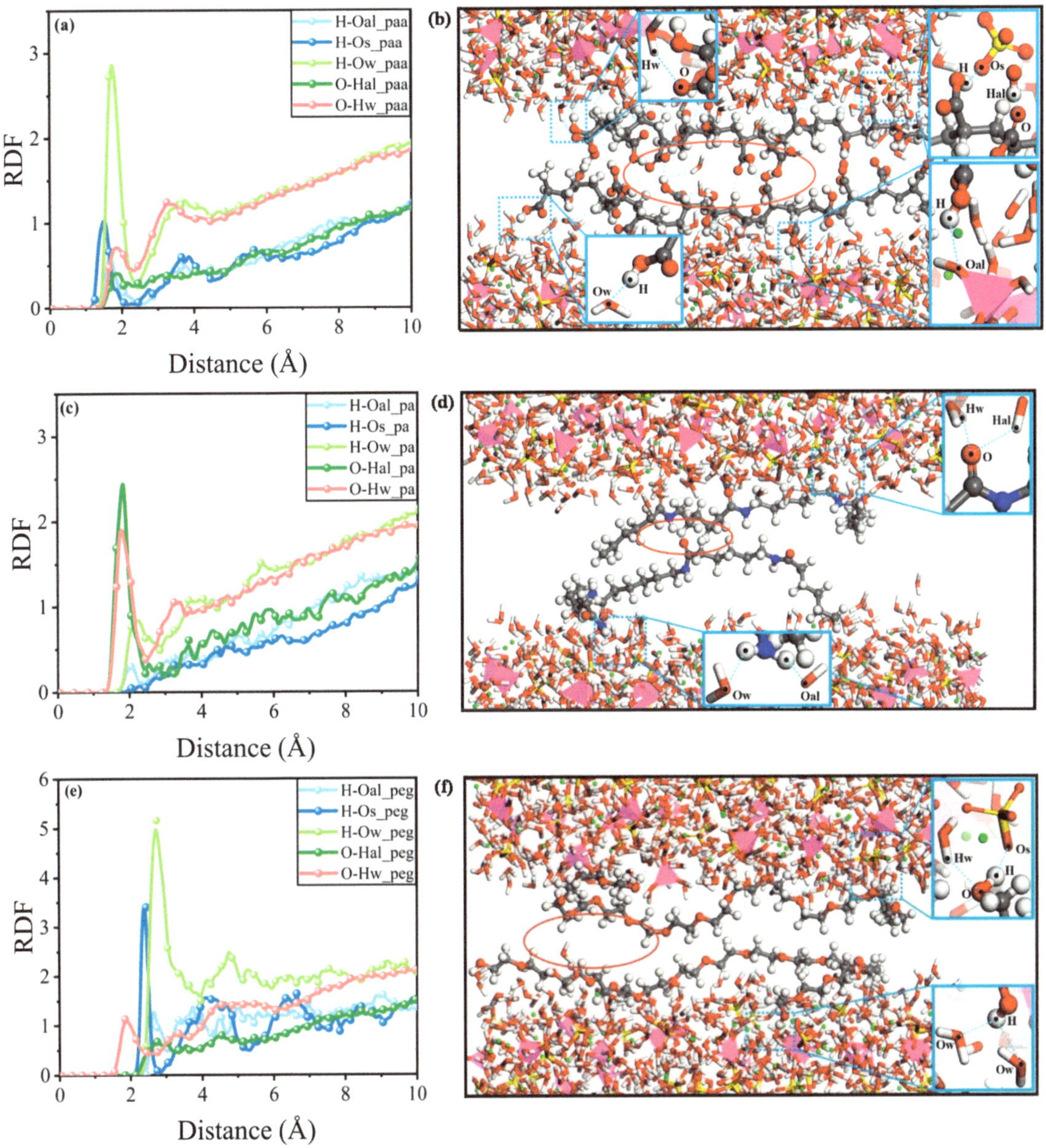

Figure 7. (**a**,**c**,**e**) represent RDF graphs of hydrogen bonds between Ettringite and PAA, PA and PEG, respectively. (**b**,**d**,**f**) represent the structure diagram of hydrogen bonds between Ettringite and PAA, PA and PEG, respectively.The inside of the red circle represents the hydrogen bond between the polymers. The inside of the blue circle represents the hydrogen bond between the polymer and the ettringite.

The H-bond RDF curve of PEG is shown in Figure 7e. It can be seen from the curve that PEG mainly forms H bonds with water molecules and sulfate, while the H bond with the aluminum hydroxyl group is weak, and its bond is shown in the blue box in Figure 7f By comparing the left of Figure 7b,d,f, it can be found that the H-bond relationship between molecules is PAA > PA > PEG, and the intermolecular interaction of polymers enhances the interfacial bonding properties between polymers and ettringite substrate.

3.4.3. Time Correlation Function

The bond formation and coordination number of different polymers and ettringite substrates were analyzed in the previous section, and bond stability also determines the strength of bond performance. The time correlation function (TCF) can be used to describe the stability of chemical bonds between different ions; the faster the TCF curve decays with time, the weaker the bond connection [42]. Figure 8a is the time correlation function of the Ca–O ionic pair. It can be seen from the figure that the TCF value of the Ca–O ionic pair of PAA is maintained at about 0.9, and the stability is the best. The two Ca–O ionic pairs of PEG exhibit different stability, and the bonding stability of oxygen atoms in the main chain is higher than that of the terminal hydroxyl group. It can be shown that the main-chain oxygen atom is the main source of PEG forming the Ca–O ionic pair. The Ca–O ionic pair stability of PA is between PAA and PEG. In analyzing the stability order of the Ca–O ionic pair among three polymers—PAA, PA and PEG—it is evident that PAA exhibits the highest stability, followed by PA and PEG. This order aligns consistently with prior examinations of bond formation. The intricate interplay of chemical structures within each polymer contributes to the observed variations in Ca–O ionic pair stability. The preference for PAA suggests specific molecular features enhancing its bond strength, reinforcing the importance of understanding polymer characteristics in the context of bond dynamics. The established order sheds light on the nuanced relationships between polymer composition and the strength of Ca–O ionic pairs, offering valuable insights for applications in various scientific and industrial domains reliant on these materials.

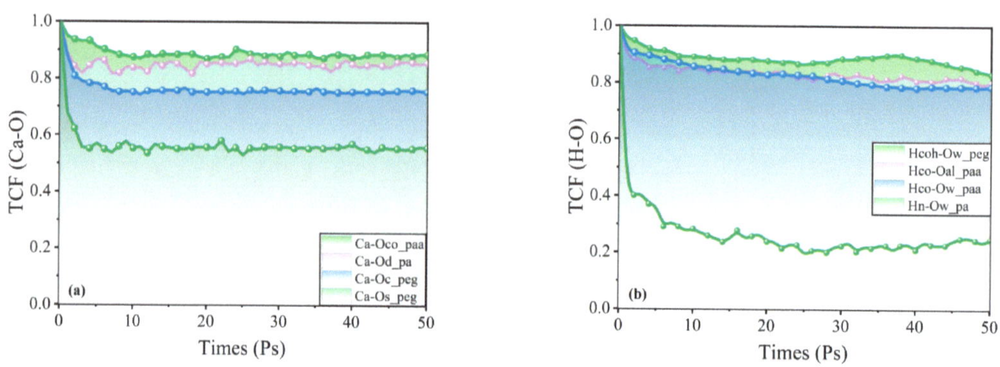

Figure 8. Time correlation function of ettringite and PAA, PA, PEG connection: (**a**) Ca–O, (**b**) H bond.

Figure 8b shows the stability of the H bond. The stability of the H bond formed by the amino group of PEG and ettringite is relatively poor, and the TCF value is about 0.25. For PAA and PA, although the above local structure analysis results show that the number of H bonds formed is small, its stability is high, which is inseparable from the fact that there are more water layers on the surface of the ettringite substrate. The H bond between polymer and substrate is still an important source of interfacial bond performance.

3.5. Dynamic Coordination Number

According to the above microscopic analysis of local structure, the bond performance between polymer and ettringite interface mainly depends on the interaction of Ca–O ionic pair strength and H-bond interaction. In order to quantitatively characterize the changes in the interface during the tensile process, the changes in coordination number over time, i.e., dynamic coordination numbers, of the two bonds were statistically analyzed, as shown in Figure 9. As can be seen from Figure 9a in the Ca–O dynamic coordination map, the curve shows a trend of rising first and is then stable in the first 300 ps, which is consistent with the equilibrium stage of polymer and substrate in the first 300 ps, where the two interact and the coordination number increases first and then achieves stable adsorption. At the stretching stage of 300 ps, the coordination number showed a downward trend. At a1, a2 and a3, the coordination number of the corresponding polymer decreased sharply and then remained stable. Combined with Figure 3, it can be found that this turning point corresponds to the time when the polymer separates from the substrate, which can reverse-prove that the three polymers bond with the substrate through Ca–O ionic pairing. In addition, the coordination number relationship of the three polymers can also be the strength of the bond: PAA > PA > PEG. Similarly, the dynamic coordination number of H bonds (Figure 9b) also shows a corresponding sharp decrease and stable stage at the time of polymer separation, b1 and b2, respectively, but peg does not change significantly due to the small number of H bonds, which is consistent with the previous analysis. The H bond is not the main source of PEG bond performance.

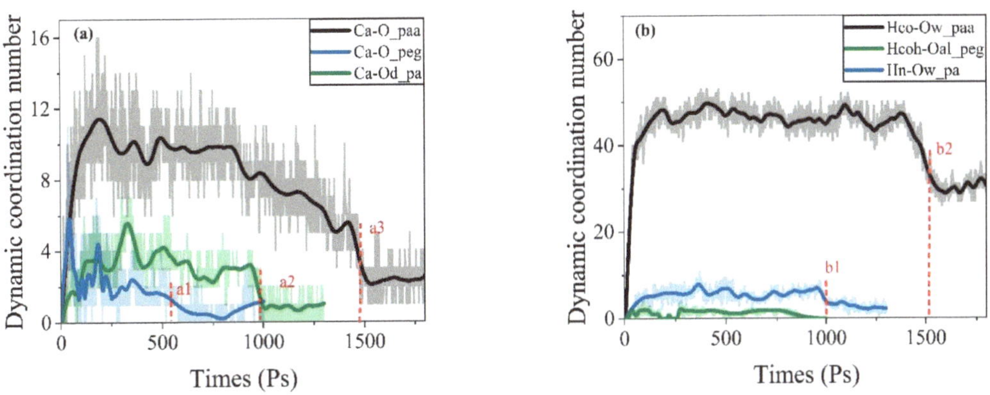

Figure 9. Dynamic coordination number: (**a**) Ca–O, (**b**) H bond. a1, a2, and a3 represent the coordination number of Ca–O bonds at 500 ps, 1000 ps, and 1500 ps respectively. b1 and b2 represent the coordination number of hydrogen bonds at 1000 ps and 1500 ps, respectively.

4. Conclusions

In this work, the interfacial bonding properties between ettringite and polymers were investigated by molecular dynamics for the first time. The separation process of four polymers, PA, PAA, PEG and PP, from ettringite was simulated by steered molecular dynamics to characterize the strength of the interfacial bonding properties of the four polymers, and the microscopic mechanisms of the differences in interfacial bonding properties were analyzed by local structural and dynamic characterization. The main findings of this work are as follows:

1. Under the tensile dynamics simulation adopted in this paper, the order of tension required for the separation of the four polymers from the interface bond is PAA > PA > PEG > PP. In the interaction with polar ettringite substrate, three polar polymers show better bonding properties than non-polar PP, and the types of polar functional groups in the polymer have a key influence on the strength of the interaction.
2. Through the microscopic mechanism analysis of local structure, it is found that the polymer can form a Ca–O ionic pair with ettringite surface calcium atoms, resulting in a strong interaction. Oxygen atoms on PA, PAA and PEG non-backbone chains can provide sites to bind calcium ions, and oxygen atoms on the PEG backbone

chain can also provide sites for calcium atoms, but, due to limitations in freedom, the coordination number value is relatively small.

3. In addition to the Ca–O ionic pair, the H bond between ettringite and polymer also determines the bond performance of the interface. The carboxyl group of PAA, the amino group of PA and the hydroxyl group of PEG provide sites for the formation of H bonds.

This study unveils the molecular-scale interface bond mechanism between ettringite and polymers, providing valuable insights into characterizing the bond properties of four types of polymers on ettringite surface. These findings offer guidance for advancing polymer-modified sulfoaluminate-cement-based materials.

Author Contributions: Methodology, H.Z.; Investigation, L.Z., H.Z. and H.X.; Writing—original draft, L.Z. All authors have read and agreed to the published version of the manuscript.

Funding: This research received no external funding.

Data Availability Statement: Data are contained within the article.

Conflicts of Interest: The authors declare no conflict of interest.

References

1. Taylor, M.; Tam, C.; Gielen, D. Energy efficiency and CO_2 emissions from the global cement industry. *Korea* **2006**, *50*, 61–67.
2. Sakai, E.; Sugita, J. Composite mechanism of polymer modified cement. *Cem. Concr. Res.* **1995**, *25*, 127–135. [CrossRef]
3. Wang, R.; Li, J.; Zhang, T.; Czarnecki, L. Chemical interaction between polymer and cement in polymer-cement concrete. *Bull. Pol. Acad. Sciences. Tech. Sci.* **2016**, *64*, 785–792. [CrossRef]
4. Song, S.H.; Liu, F.T.; Huang, Y.B. Effect of Polymer Latex and Fiber on Properties of Sulpho Aluminate Cement Mortar. *Adv. Mater. Res. Trans. Tech. Publ.* **2012**, *450*, 402–406. [CrossRef]
5. Nematzadeh, M.; Dashti, J.; Ganjavi, B. Optimizing compressive behavior of concrete containing fine recycled refractory brick aggregate together with calcium aluminate cement and polyvinyl alcohol fibers exposed to acidic environment. *Constr. Build. Mater.* **2018**, *164*, 837–849. [CrossRef]
6. Zhang, C.; Liu, Y.; Zhang, M. PP/PVA Fiber Reinforced Sulphoaluminate Cement-Based Rapid Repair Material. *Bull. Chin. Ceram. Soc.* **2021**, *40*, 2174–2183. [CrossRef]
7. Xue-jun, X.; Yu, L.; Yu-fei, J. Experimrntal Study on Mechanical Performances of Fiber Reinforced Repair. *China Concr. Cem. Prod.* **2018**, *12*, 49–52. [CrossRef]
8. Zhang, H.; Zhou, R.; Liu, S.; Zhu, Y.; Wang, S.; Wang, J.; Guan, X. Enhanced toughness of ultra-fine sulphoaluminate cement-based hybrid grouting materials by incorporating in-situ polymerization of acrylamide. *Constr. Build. Mater.* **2021**, *292*, 123421. [CrossRef]
9. Tian, Y.; Li, Z.J.; Ma, H.Y.; Jin, X.Y.; Jin, N.G. Physical and chemical influence of polyacrylate latex on cement mortars. *Adv. Mater. Res. Trans. Tech. Publ.* **2011**, *261*, 807–811. [CrossRef]
10. Liu, H.; Zhao, P.; Lu, L.; Wang, S.; Chen, M.; Cheng, X. Effect of Composition on Waterproofing Property of Silicone Modified Polyacrylate Emulsion-Sulphoaluminate Cement Composite Coating: Optimization by Response Surface Methodology. *J. Inorg. Organomet. Polym. Mater.* **2019**, *29*, 429–435. [CrossRef]
11. Wang, M.; Wang, R.; Zheng, S.; Farhan, S.; Yao, H.; Jiang, H. Research on the chemical mechanism in the polyacrylate latex modified cement system. *Cem. Concr. Res.* **2015**, *76*, 62–69. [CrossRef]
12. Rapaport, D.C.; Rapaport, D.C.R. *The Art of Molecular Dynamics Simulation*; Cambridge University Press: Cambridge, UK, 2004.
13. Alder, B.J.; Wainwright, T.E. Studies in molecular dynamics. I. General method. *J. Chem. Phys.* **1959**, *31*, 459–466. [CrossRef]
14. Hansson, T.; Oostenbrink, C.; van Gunsteren, W. Molecular dynamics simulations. *Curr. Opin. Struct. Biol.* **2002**, *12*, 190–196. [CrossRef] [PubMed]
15. Hou, D.; Yu, J.; Wang, P. Molecular dynamics modeling of the structure, dynamics, energetics and mechanical properties of cement-polymer nanocomposite. *Compos. Part B Eng.* **2019**, *162*, 433–444. [CrossRef]
16. Wang, P.; Qiao, G.; Zhang, Y.; Hou, D.; Zhang, J.; Wang, M.; Wang, X.; Hu, X. Molecular dynamics simulation study on interfacial shear strength between calcium-silicate-hydrate and polymer fibers. *Constr. Build. Mater.* **2020**, *257*, 119557. [CrossRef]
17. Qiao, G.; Hou, D.; Wang, P.; Lu, Z. Insights on failure modes of calcium-silicate-hydrate interface strengthened by polyacrylamides: Structure, dynamic and mechanical properties. *Constr. Build. Mater.* **2021**, *278*, 122406. [CrossRef]
18. Liu, K.; Cheng, X.; Ma, Y.; Gao, X.; Zhang, C.; Li, Z.; Zhuang, J. Analysis of interfacial nanostructure and interaction mechanisms between cellulose fibres and calcium silicate hydrates using experimental and molecular dynamics simulation data. *Appl. Surf. Sci.* **2020**, *506*, 144914. [CrossRef]
19. Zhou, Y.; Tang, L.; Liu, J.; Miao, C. Interaction mechanisms between organic and inorganic phases in calcium silicate hydrates/poly (vinyl alcohol) composites. *Cem. Concr. Res.* **2019**, *125*, 105891. [CrossRef]

20. Zhou, Y.; Hou, D.; Jiang, J.; She, W.; Li, J. Molecular dynamics study of solvated aniline and ethylene glycol monomers confined in calcium silicate nanochannels: A case study of tobermorite. *Phys. Chem. Chem. Phys.* **2017**, *19*, 15145–15159. [CrossRef]
21. Duan, Y.; Zheng, H.; Wang, P.; Hou, D.; Wang, M.; Yin, B.; Li, S. Molecular dynamics simulation study on the hydrophobic mechanism of Ettringite nanoporous channels modified by silane and silane/graphene oxide. *Appl. Surf. Sci.* **2023**, *623*, 156975. [CrossRef]
22. Moore, A.; Taylor, H. Crystal structure of ettringite. *Acta Crystallogr. Sect. B Struct. Crystallogr. Cryst. Chem.* **1970**, *26*, 386–393. [CrossRef]
23. Jueshi, Q.; Jinchen, Y.; Huaqiang, S.; Ying, M. Formation and Function of Ettringite in Cement Hydrates. *J. Chin. Ceram. Soc.* **2017**, *45*, 1569–1581. [CrossRef]
24. Skoblinskaya, N.; Krasilnikov, K. Changes in crystal structure of ettringite on dehydration. 1. *Cem. Concr. Res.* **1975**, *5*, 381–393. [CrossRef]
25. Cygan, R.T.; Liang, J.-J.; Kalinichev, A.G. Molecular models of hydroxide, oxyhydroxide, and clay phases and the development of a general force field. *J. Phys. Chem. B* **2004**, *108*, 1255–1266. [CrossRef]
26. Honorio, T.; Guerra, P.; Bourdot, A. Molecular simulation of the structure and elastic properties of ettringite and monosulfoaluminate. *Cem. Concr. Res.* **2020**, *135*, 106126. [CrossRef]
27. Cannon, W.R.; Pettitt, B.M.; McCammon, J.A. Sulfate anion in water: Model structural, thermodynamic, and dynamic properties. *J. Phys. Chem.* **1994**, *98*, 6225–6230. [CrossRef]
28. Shahriyari, R.; Khosravi, A.; Ahmadzadeh, A. Nanoscale simulation of Na-Montmorillonite hydrate under basin conditions, application of CLAYFF force field in parallel GCMC. *Mol. Phys.* **2013**, *111*, 3156–3167. [CrossRef]
29. Plimpton, S.; Crozier, P.; Thompson, A. LAMMPS-large-scale atomic/molecular massively parallel simulator. *Sandia Natl. Lab.* **2007**, *18*, 43.
30. Nosé, S. A unified formulation of the constant temperature molecular dynamics methods. *J. Chem. Phys.* **1984**, *81*, 511–519. [CrossRef]
31. Keil, F. Multiscale modelling in computational heterogeneous catalysis. In *Multiscale Molecular Methods in Applied Chemistry*; Springer: Berlin/Heidelberg, Germany, 2011; pp. 69–107.
32. Evans, D.J.; Hoover, W.G.; Failor, B.H.; Moran, B.; Ladd, A.J. Nonequilibrium molecular dynamics via Gauss's principle of least constraint. *Phys. Rev. A* **1983**, *28*, 1016. [CrossRef]
33. Wang, P.; Yang, Q.; Jin, Z.; Hou, D.; Wang, M. Effects of water and ions on bonding behavior between epoxy and hydrated calcium silicate: A molecular dynamics simulation study. *J. Mater. Sci.* **2021**, *56*, 16475–16490. [CrossRef]
34. Bezerra, E.M.; Joaquim, A.P.; Savastano, H. Some properties of fiber-cement composites with selected fibers. In Proceedings of the Conferencia Brasileira de Materiais e Tecnologias Não Convencionais: Habitações e Infra-Estrutura de Interesse Social Bra-sil-NOCMAT, Pirassununga, SP, Brasil, 3 November 2004; pp. 34–43.
35. Humphrey, W.; Dalke, A.; Schulten, K. VMD: Visual molecular dynamics. *J. Mol. Graph.* **1996**, *14*, 33–38. [CrossRef] [PubMed]
36. Han, Q.; Yang, Y.; Zhang, J.; Yu, J.; Hou, D.; Dong, B.; Ma, H. Insights into the interfacial strengthening mechanism of waste rubber/cement paste using polyvinyl alcohol: Experimental and molecular dynamics study. *Cem. Concr. Compos.* **2020**, *114*, 103791. [CrossRef]
37. Taylor, H.; Famy, C.; Scrivener, K. Delayed ettringite formation. *Cem. Concr. Res.* **2001**, *31*, 683–693. [CrossRef]
38. Wang, M. *Reaserch on the Properties and Mechanism of High Performance Cement–Based Materials*; Northwestern Polytechnical University: Xi'an, China, 2018.
39. Sowoidnich, T.; Rachowski, T.; Rößler, C.; Völkel, A.; Ludwig, H.-M. Calcium complexation and cluster formation as principal modes of action of polymers used as superplasticizer in cement systems. *Cem. Concr. Res.* **2015**, *73*, 42–50. [CrossRef]
40. Zhou, Y.; Huang, J.; Yang, X.; Dong, Y.; Feng, T.; Liu, J. Enhancing the PVA fiber-matrix interface properties in ultra high performance concrete: An experimental and molecular dynamics study. *Constr. Build. Mater.* **2021**, *285*, 122862. [CrossRef]
41. Chen, B.; Qiao, G.; Hou, D.; Wang, M.; Li, Z. Cement-based material modified by in-situ polymerization: From experiments to molecular dynamics investigation. *Compos. Part B Eng.* **2020**, *194*, 108036. [CrossRef]
42. Hou, D.; Li, Z.; Zhao, T. Reactive force field simulation on polymerization and hydrolytic reactions in calcium aluminate silicate hydrate (C–A–S–H) gel: Structure, dynamics and mechanical properties. *Rsc. Adv.* **2015**, *5*, 448–461. [CrossRef]

buildings

MDPI

Article

Effect of Citric Acid-Modified Chitosan on Hydration Regulation and Mechanism of Composite Cementitious Material Syste

Liguo Wang [1], Zhibin Qin [1], Jiandong Wu [2], Guangxia Sheng [1], Han Wang [1], Kai Liu [3], Xiaobin Dong [1], Fengjuan Wang [1] and Jinyang Jiang [1,*]

1 School of Materials Science and Engineering, Southeast University, Nanjing 211189, China; wlg_seu@sina.com (L.W.); 13834133171@zt17.cn (Z.Q.); shenggx0812@163.com (G.S.); hanwangq@gmail.com (H.W.); 220212321@seu.edu.cn (X.D.); fengjuan19921118@sina.com (F.W.)

2 Shandong Provincial Communications Planning and Design Institute Group Co., Ltd., Jinan 250101, China; jdwu_mail@163.com

3 Jiangsu China Construction Ready Mixed Concrete Co., Ltd., NanJing 210033, China; lk230131@126.com

* Correspondence: jiangjinyang16@163.com

Abstract: The temperature stress caused by the large temperature difference is the main factor causing harmful cracks in large-volume concrete. The introduction of admixtures is beneficial to reduce the temperature difference inside and outside the large-volume concrete. This study investigated the mechanism of how citric acid-modified chitosan (CAMC) affects the hydration heat release process and hydration products of composite cementitious materials. Through methods such as hydration heat, X-ray diffraction (XRD), mercury intrusion porosimetry (MIP), scanning electron microscopy (SEM), and nuclear magnetic resonance (NMR), the mechanism of how CAMC controls the hydration heat release process and hydration products of composite cementitious materials was revealed. The results show that the addition of CAMC delayed the hydration process of cementitious materials without affecting the type of hydration products but affected the content of each phase of hydration products. As the hydration process proceeded, the total porosity of all samples decreased, the volume of large pores decreased, and the volume of small pores increased. As the content of CAMC increased, the pore diameter of the hardened paste gradually became smaller, the proportion of large pores decreased, and the later hydration microstructure became more dense. The increase in CAMC dosage resulted in a decrease in the peak intensity of Q^2 in the paste, indicating that Al atoms in Q^2 (^{1}Al) existed in the form of Alcoordination, which proves that CAMC reduced its hydration degree and delayed cement hydration.

Keywords: CAMC; composite cementitious materials; hydration exothermic process; hydration products

Citation: Wang, L.; Qin, Z.; Wu, J.; Sheng, G.; Wang, H.; Liu, K.; Dong, X.; Wang, F.; Jiang, J. Effect of Citric Acid-Modified Chitosan on Hydration Regulation and Mechanism of Composite Cementitious Material System. *Buildings* **2024**, *14*, 41. https://doi.org/10.3390/buildings14010041

Academic Editor: Denny Coffetti

Received: 7 November 2023
Revised: 23 November 2023
Accepted: 18 December 2023
Published: 22 December 2023

1. Introduction

Concrete, as a porous composite material, is widely used worldwide due to its convenient material selection, excellent mechanical properties, and strong durability [1–5]. Concrete has become an indispensable building material for the construction of major military and civil infrastructure projects. However, in large volume concrete structures, due to the large volume of the structure, a significant proportion of cement content is essential. In the early construction process, cement releases a large amount of hydration heat due to hydration reactions, leading to a rapid increase in the internal temperature of concrete. Due to the large volume of poured concrete, this heat cannot be dissipated in a timely manner. On the contrary, when the concrete surface comes into contact with the external environment, it dissipates heat quickly. Therefore, the internal and external thermal expansion of mass concrete are uneven, resulting in huge temperature stress. When the temperature stress exceeds the ultimate tensile strength of concrete, the cracks caused are called temperature cracks [6]. The cracking problem caused by temperature stress in large-volume concrete structures is quite common, accounting for about 80% of engineering structural cracks, seriously affecting the durability of concrete, leading to early retirement of buildings, causing huge losses to economic construction, and greatly increasing maintenance costs. Therefore, how to reasonably and effectively reduce the hydration heat of large-volume concrete is of great engineering significance for large buildings.

Ismail et al. [7] studied the effect of vinyl acetate wastewater (generated during the coating production process) on the initial setting time and hydration heat of concrete. The results showed that adding vinyl acetate wastewater to concrete can not only delay the setting time of concrete but also reduce the total temperature rise of concrete. Khil et al. [8] studied the application of phase change materials (PCM) in mass concrete mixtures and believed that the impact of PCM on the strength of mass concrete can be ignored. The use of PCM can effectively reduce adiabatic temperature rise and temperature stress. Choi et al. [9] evaluated the feasibility of phase change materials (PCM) in suppressing the hydration heat of large-volume concrete through experimental tests (micro conductivity, simplified adiabatic temperature rise, heating, and compressive strength tests). The experimental results showed that PCM has good latent heat characteristics, which help to prevent volume changes and microcracks caused by thermal stress in large-volume concrete. Organic additives can affect the nucleation mechanism of C-S-H in mass concrete, thereby regulating the hydration heat of mass concrete in a reasonable manner [10,11]. Chitosan and its derivatives have been introduced as a new type of admixture into concrete materials due to their excellent hydrophilicity and biocompatibility. Wang et al. [12] modified chitosan by citric acid and found that the modified chitosan can regulate the exothermic process in the cement hydration process. However, the influence of modified chitosan on the hydration process and formation mechanism of hydration products of cementitious materials in large-volume concrete still deserves further research.

Therefore, this study focuses on the cement fly ash slag cementitious material system as the main composite cementitious material and systematically studies the influence of modified chitosan on the hydration process and formation mechanism of hydration products of the composite cementitious material.

2. Materials and Methods

2.1. Materials

The cementitious material is P•II 52.5 Portland cement (PC), and the auxiliary cementitious materials are special-grade fly ash (FA) and S95-grade blast furnace granulated slag (GGBS). The specific chemical composition, mineral composition, and particle size distribution are shown in Table 1 and Figure 1, respectively. The main physical parameters of Portland cement are shown in Table 2. The low-temperature rising polymer used in the experiment is citric acid-modified chitosan (CAMC), and the modification method refers to the literature [12–14].

Table 1. Chemical and mineral compositions of cementitious materials (wt%) [12].

Material	CaO	SiO_2	Al_2O_3	Fe_2O_3	SO_3	MgO	K_2O	Na_2O	TiO_2	Loss
PC	63.62	19.70	4.45	2.93	2.93	1.28	0.68	0.12	0.27	3.92
FA	17.60	65.67	6.84	0.06	-	0.08	0.04	0.035	0.015	9.639
GGBS	40.918	34.125	15.921	0.53	2.31	4.60	0.218	0.364	0.695	0.319

Table 2. Physical properties of Portland cement.

Density (g/cm^3)	Specific Surface Area (m^2/kg)	Water Demand (wt.%)	Initial Setting Time (min)	Final Setting Time (min)	Flexural Strength (MPa)		Compressive Strength (MPa)	
					3 d	28 d	3 d	28 d
3.12	372	30	174	275	5.10	8.15	30.75	54.04

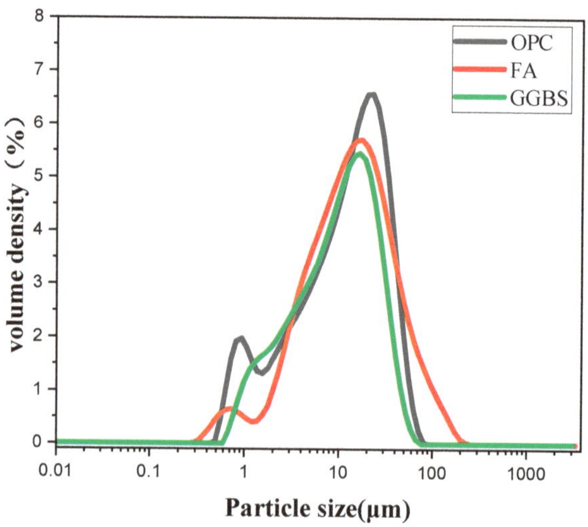

Figure 1. Particle size distribution of cementing material.

2.2. Specimen Preparation

The experimental coordination is shown in Table 3. The clean slurry mixer is used for mixing. Firstly, cement, fly ash, mineral powder, and CAMC are added to the mixer for dry mixing for 2 min. Then, wet mix for 3 min to achieve good workability of the material. After the sample is formed, it is molded and demolded after 1 d of standard curing. After demolding the specimen, transfer it to a standard curing room (temperature (20 ± 1 °C), relative humidity ≥ 95%) for curing.

Table 3. Hydration heat mix ratio of composite cementitious materials with different contents of low-temperature liter polymer (kg/m^3).

NO.	PC	FA	GGBS	CAMC	W
CA1	340	110	60	0.1%	204
CA2	340	110	60	0.2%	204
CA3	340	110	60	0.3%	204

Note: CA1 represents 0.1% citric acid-modified chitosan (CAMC).

2.3. Tests for Property Evaluation

2.3.1. Heat of Hydration

The experiment used an isothermal calorimeter produced by New Castle DE in the United States to measure the heat release rate and heat release of the composite cementitious material after 120 h of hydration. In this article, deionized water is used as a comparison sample, with a test temperature of 20 °C and a temperature-controlled room as the test environment to ensure the stability of the reference line during instrument calibration. Moreover, after the cementitious material is evenly mixed with water, it is immediately placed in the channel of an isothermal calorimeter, and the hydration heat release rate and total heat release of the composite cementitious material can be continuously measured.

2.3.2. Phase Analysis

Crush the sample cured to a fixed age and soak it in alcohol for 3 days to terminate hydration. Dry the sample at 60 °C, grind it into powder, and pass it through a 200-mesh sieve. The D8-ADVANCED Diffractometer of Germany Bruker company (Saarbrucken, Germany) was used for testing. The tube voltage was 40 kv, the tube current was 40 mA, the Cu was the target, the scanning range was 5°~80°, the scanning speed was 2°/min, and the step size was 0.02°.

The internal standard method is used for diffraction for quantitative testing. Corundum was selected as a standard sample, and Topas 3-C software was used for analysis [12]. Before conducting the test, mix 0.5 g of corundum and 2 g of sample powder in a centrifuge tube for 5 min and then place them on a rotary table for 2 h to ensure uniform mixing.

2.3.3. MIP

The IV9510 mercury porosimeter (Mike Instruments, GA, USA) is used to test the internal pore structure of hardened specimens. The high pressure of the equipment can reach 60,000 psi, and the theoretically measurable minimum aperture is 3.2 nm. To save mercury solution and improve the accuracy of the results, crushed samples that have terminated hydration and been fully dried should be filled with the expansion bottle as much as possible.

2.3.4. SEM

The scanning electron microscope is an FEI 3D field emission environment scanning electron microscope produced by FEI Company (Hillsborough, OR, USA), which is used to observe the microscopic morphology of the sample. Scanning the sample using a focused electron beam (energy level order of keV) with a resolution of 3 nm~6 nm. When observing the microstructure of the hydration sample, cut off the middle part of the hydration sample and crush it. After the hydration is terminated with alcohol, dry it at 65 °C and select a flat drying sample to spray gold for observation.

2.3.5. NMR

The Bruker AVANCE III 400 WB superconducting nuclear magnetic resonance spectrometer was used to test the ^{29}Si and ^{27}Al nuclear magnetic resonance spectra of powder samples. In the experiment, a 4 mm zirconia rotor was used for testing, and the resonant frequencies of ^{29}Si and ^{27}Al were 79.51 and 104.29 MHz, respectively. For the ^{29}Si device, the magic angle rotation rate is 8 kHz, and the pulse width is 4.0 μs, with a cycle delay time of 10 s. Tetramethylsilane (TMS) was used as a reference standard. For ^{27}Al, the magic angle rotation rate is 10 kHz, and the pulse width is 1.5 μs, with a cycle delay time of 2 s, using 1.0 mol/L $Al(NO_3)_3$ as the reference sample.

3. Result and Discussion

3.1. Heat of Hydration

The heat release rate curve and cumulative water accumulation heat release curve of CAMC with different dosages on composite cementitious materials with curing age are shown in Figure 2. It can be seen that the addition of CAMC has a significant impact on the hydration heat release rate of the composite cementitious system. With the increase of CAMC dosage, both the first and second hydration heat release peaks of the slurry exhibit a delayed phenomenon, and the peak intensity significantly decreases. Compared to the CA1 sample, the maximum heat release rate at the first exothermic peak of the CA3 sample decreased by about 40%. Moreover, the maximum exothermic peak of the CA3 sample was delayed by about 13 h compared to the CA1 sample, which significantly prolonged the induction period. It indicates that the introduction of CAMC can delay the hydration process of cement particles and significantly reduce the heat release rate of slurry hydration, which is of great significance for the hydration reaction of cementitious materials in large-volume concrete. Figure 2b shows that in the early stage of slurry hydration, CAMC can reduce the accumulated heat release of the slurry, which is related to its delay in cement hydration. However, the impact of CAMC on the total heat release of the slurry is relatively small, which may depend on the amount of CAMC added. In summary, the introduction of CAMC into cement slurry reduces the hydration rate of cement without affecting the total heat release of the slurry, preventing a large accumulation of hydration heat, which is beneficial for reducing the temperature stress and cracking risk of concrete.

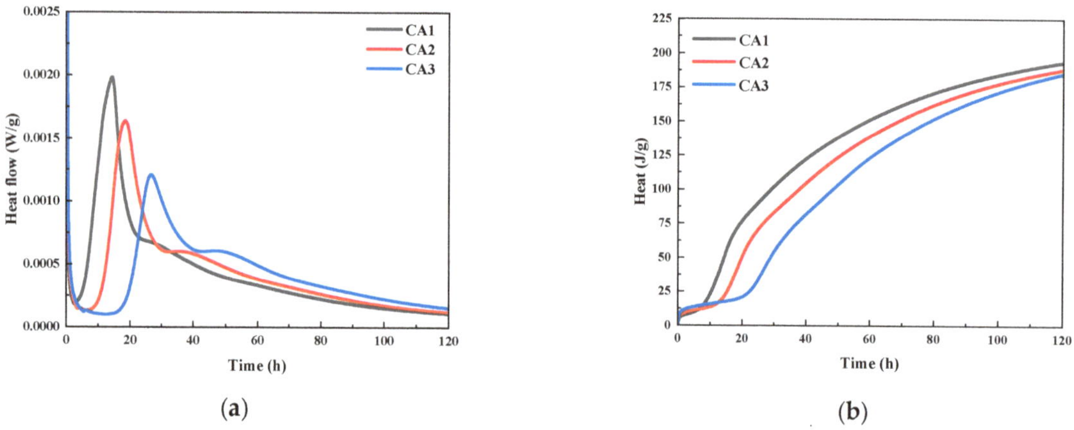

Figure 2. Effect of CAMC on hydration heat of composite cementitious materials: (**a**) heat flow; (**b**) cumulative heat release.

3.2. Phase Analysis

The XRD spectra of samples were measured at different ages to dynamically track the hydration products of composite cementitious materials. The test results are shown in Figure 3.

Figure 3a–c shows that with the increase of hydration age from 3 h to 3 d, the peak values of C_2S and C_3S of all samples gradually decrease, and the peak values of calcium hydroxide (CH) gradually increase, which is the inevitable result of hydration of cementitious materials. It can be seen from Figure 3a that the peak intensity of the CH characteristic peak (18.1°) of all samples at 3 h is very weak and almost unobservable on the graph. After 24 h of hydration, a significant CH peak appeared in the CA2 sample, and the peak intensity was close to that of the CA1 sample. The CH characteristic peak intensity of the CA3 sample was still extremely low. Therefore, the introduction of CAMC significantly delayed the formation time of CH. It can be seen from Figure 3c that the CH characteristic peaks of all samples are very obvious. As the CAMC dosage increases from 0.1% to 0.3%, the CH peak intensity gradually increases. However, the strength of C_2S and C_3S peaks decreased, because the hydration rate of the CA1 sample was the least inhibited, and the generated CH participated in the reaction to generate hydration products such as C-S-H gel. As the curing age increases, the CH characteristic peaks of all samples show a gradual upward trend. This is consistent with the phenomenon in the hydration heat release rate curve of the slurry in Figure 3a (after about 40 h of hydration, the higher the CAMC content, the faster the hydration heat release rate of the slurry), indicating that CAMC mainly plays a role in delaying cement hydration in the early stage, and then C_2S and C_3S delayed by CAMC begin to hydrate. Therefore, there is a phenomenon that the CH content of hydration products increases with the increase of CAMC dosage during the 3–28 d age period, while the characteristic peaks of other products are less affected by CAMC dosage. In the XRD spectrum at 28 d, the characteristic peak positions of each group of slurries are almost the same, and the intensities of CH, C_2S, C_3S, AFm, and other peaks are also equivalent. This further proves that CAMC only delays cement hydration in the early stage of hydration, to some extent delaying the dissolution of C_3S, and has little impact on the total amount of hydration products. CAMC has a porous structure that makes it easier to adsorb on the surface of cement particles, thereby delaying the cement hydration process.

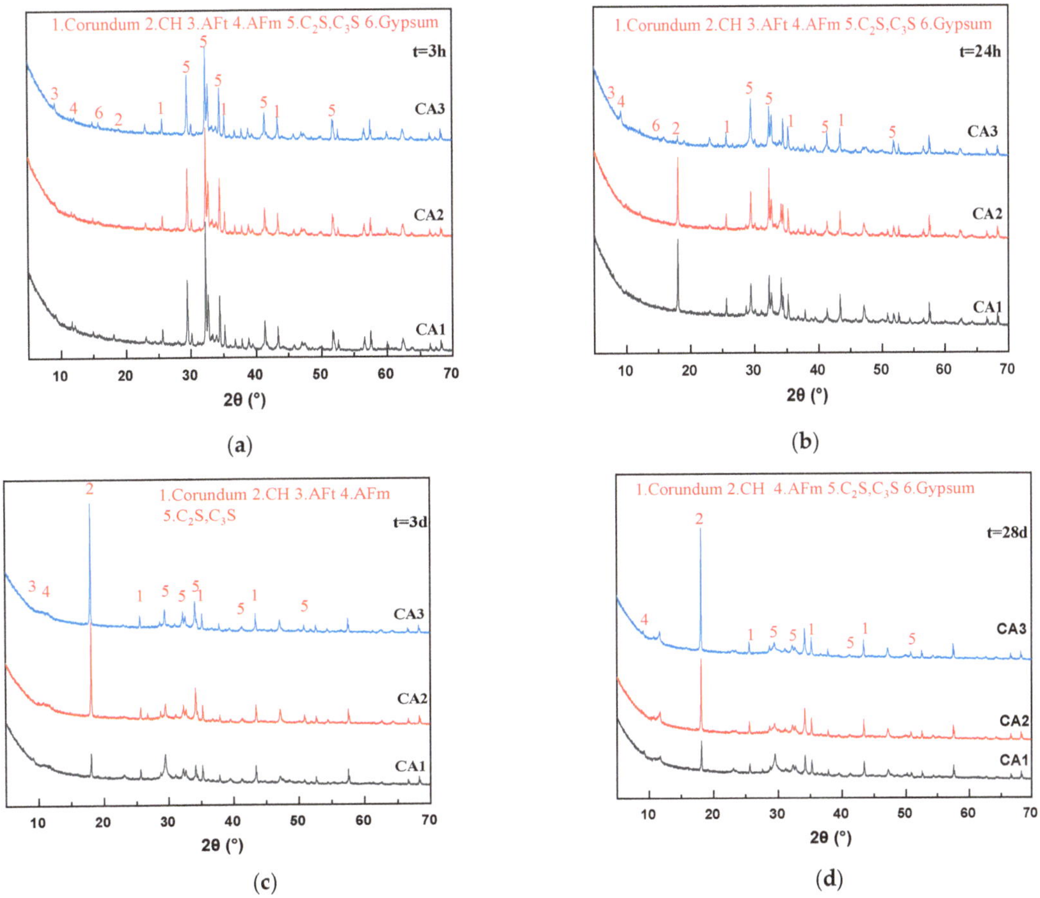

Figure 3. XRD patterns of sample at different ages. (**a**) 3 h; (**b**) 24 h; (**c**) 3 d; (**d**) 28 d.

3.3. *Degree of Hydration Reaction*

Based on the XRD data, the effect of CAMC on the reaction degree of cement clinker (C_3S, C_2S, C_3A, C_4AF) was studied. Usually, the reaction degree of cement clinker can be calculated based on the following equation.

$$(DoH)t = 1 - \frac{wc_3s(t) + wc_2s(t) + wc_3A(t) + wc_4AF(t)}{wc_3s(t_0) + wc_2s(t_0) + wc_3A(t_0) + wc_4AF(t_0)} \quad (1)$$

where $w(t)$ represents the content of C_3S, C_2S, C_3A, and C_4AF at different ages, while $w(t_0)$ represents the mass fraction of C_3S, C_2S, C_3A, and C_4AF in the un-hydrated clinker.

The influence of different amounts of CAMC on the degree of clinker hydration reaction is shown in Table 4. With the increase of CAMC content, the overall hydration rate of cement slows down and the degree of hydration reaction decreases. When the CAMC content is 0.1%, the degree of hydration reaction of cement at 0.5 d, 1 d, 3 d, 28 d, and 56 d is 19.05%, 50.41%, 59.01%, 71.06%, 73.49%, and 75.74%, respectively. When the addition of CAMC is 0.2%, the degree of cement hydration decreases to 3.32% and 27.51% at 0.5 d and 1 d, respectively. With the continuous progress of hydration reaction, the hydration degree of cement mixed with 0.1% CAMC remains the same at 3 d. However, the addition of 0.3% CAMC significantly reduced the hydration degree of cement. Compared to the CA1 sample, the hydration degree of the CA3 sample decreased by 68.7%, 21.7%, and 16.2% at 1 d, 7 d, and 56 d, respectively. By fitting the reaction degree of different clinker, the long-term reaction rate of cement hydration can be predicted based on the obtained curve. The fitting curves of different amounts of CAMC on the reaction degree of cement clinker are shown in Figure 4. By comparing the growth trend of the fitted curve, it can be seen that although adding 0.3% CAMC inhibits cement hydration in the early stage, the degree of reaction in the system will gradually increase with the increase of age.

Table 4. Pore structure of specimens with different CAMC contents.

NO.-Age	Porosity (%)	Mean Pore Diameter (nm)	Aperture Distribution Ratio (%)			
			<20 nm	20~50 nm	50~200 nm	>200 nm
CA1-3d	25.06	11.44	51.65	23.11	19.33	5.91
CA2-3d	29.42	12.41	43.88	19.29	28.70	8.13
CA3-3d	26.31	13.64	40.18	22.43	32.30	5.09
CA1-7d	21.49	8.92	67.37	23.23	3.22	6.18
CA2-7d	19.32	8.94	62.26	27.57	6.63	3.54
CA3-7d	26.83	10.94	52.79	28.64	10.15	8.42
CA1-28d	19.25	9.73	52.11	30.55	10.31	7.03
CA2-28d	20.97	9.21	61.66	26.84	4.75	6.75
CA3-28d	18.33	9.63	61.67	27.92	3.20	7.21

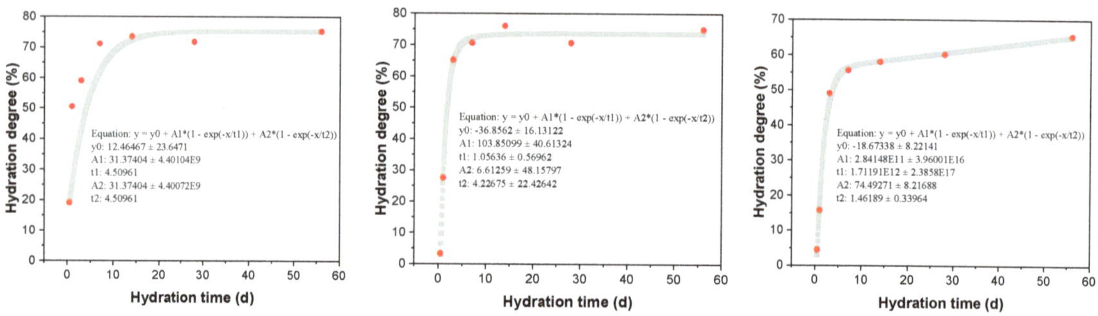

(a) (b) (c)

Figure 4. Effect of CAMC on hydration degree of cement clinker. (**a**) CA1; (**b**) CA2; (**c**) CA3.

3.4. Changes in Content of Clinker Phase and Hydration Products

The changes in the content of various clinker phases (C_3S, C_2S, C_3A, C_4AF), CH, and AFt in cement were characterized by the XRD internal standard method, and the influence of different amounts of CAMC on the phase evolution of cement was obtained, as shown in Figure 5. As the content of CAMC increases, the degree of reaction between C_3S and C_2S decreases, but the reaction of C_2S is not affected by the content of CAMC. The degree of C_3S reaction is also less affected by changes in CAMC content (the reaction degree of CA1 and CA2 samples at the same age is almost the same). Therefore, the retarding effect of CAMC is mainly on the silicon phase, while the hydration process of the aluminum phase is not affected by CAMC. This fact provides support for revealing the retarding effect of CAMC on cement. Furthermore, the mechanism of action of CAMC and $CaSO_4$ in cement retarding is different.

According to the content of CH and AFt, it can be seen that when 0.1%, CAMC is added to cement-based materials, and the CH content shows a trend of first increasing and then decreasing. This is because the content of hydration products increases with the early hydration process. However, with hydration, volcanic ash reaction occurs between fly ash and slag and CH, and CH content is consumed. As the amount of CAMC increases, the early content of CH decreases, which is caused by the retarding effect of CAMC. However, in the later stage, the CH content does not further decrease but fluctuates with hydration, while the content of fly ash and slag remain unchanged. Therefore, it has been proven that the CH content in the hydration system increases in the later stage. According to the comprehensive hydration degree of clinker, the addition of CAMC has a certain promoting effect on the hydration degree of cement in the later stage.

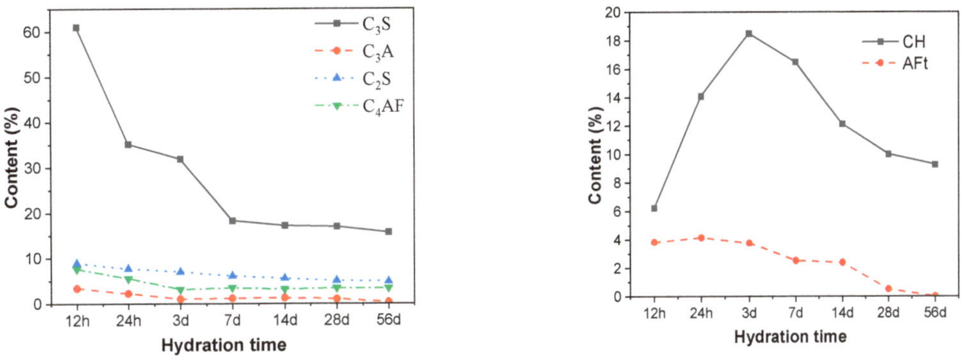

(**a**)

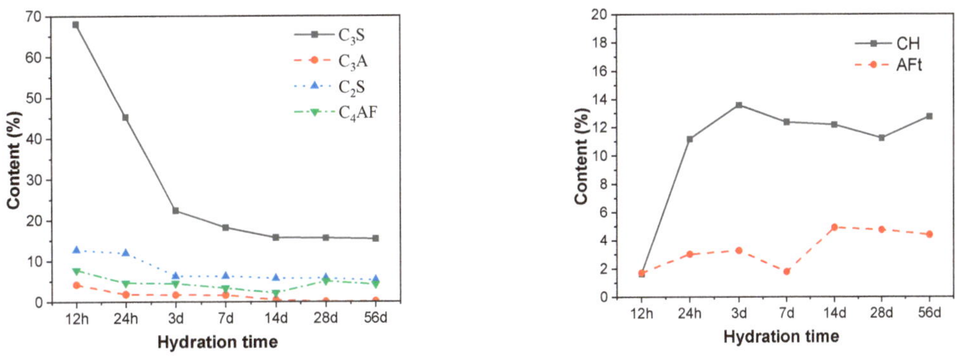

(**b**)

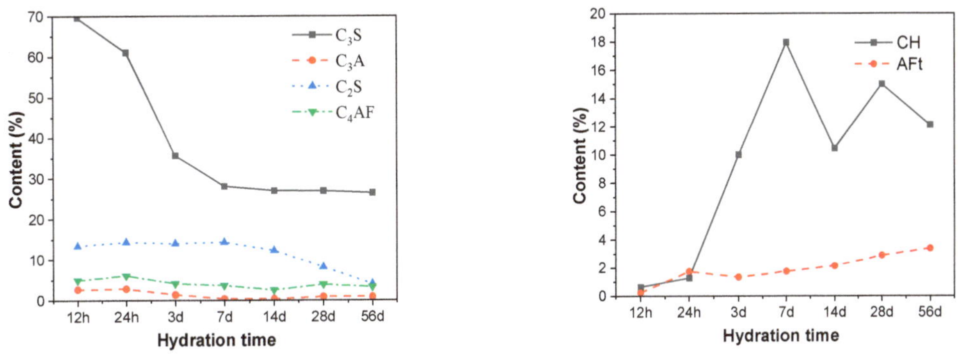

(**c**)

Figure 5. Changes in clinker phase and hydration content of different contents of CAMC slurries. (**a**) CA1; (**b**) CA2; (**c**) CA3.

3.5. MIP

Figure 6 shows the pore size distribution and integration curves of samples with different CAMC dosages after 3 d, 7 d, and 28 d of curing. It can be seen from Figure 6a,c,e that as the degree of hydration increases, the overall pore distribution curve shifts to the right, indicating a decrease in the volume of large pores and an increase in the volume of small pores. The pore structure characteristics of samples with different CAMC dosages at 3, 7, and 28 days of curing are summarized in Table 4. As the hydration age increases from 3 days to 28 days, the porosity of the CA1 sample decreases from 25.06% to 19.25%, the CA2 sample decreases from 29.42% to 20.97%, and the CA3 sample decreases from 26.31% to 18.33%, all of which demonstrate the gradual improvement of the pore structure of the hardened slurry. Early cement paste has delayed hydration, more pores, and a loose structure, which to some extent affect the pore size distribution of the cement slurry. However, as the hydration time increases, the pore structure gradually becomes denser, the total pore volume and the volume of many harmful pores gradually decrease, while the volume of harmless pores gradually increases. The structural defects caused by CAMC in the early stage will be compensated for in the later stage of hydration. Moreover, when the harmful and multi-harmful pores of the CA2 sample reach a maximum value of 3 d, and when the harmful and multi-harmful pores of the CA3 sample reach a maximum value of 7 d, but when the harmful and multi-harmful pores of the CA1 sample reach a maximum value of 28 d, this corresponds to the fact that CAMC can delay cement hydration and the delaying effect increases with the increase of CAMC content.

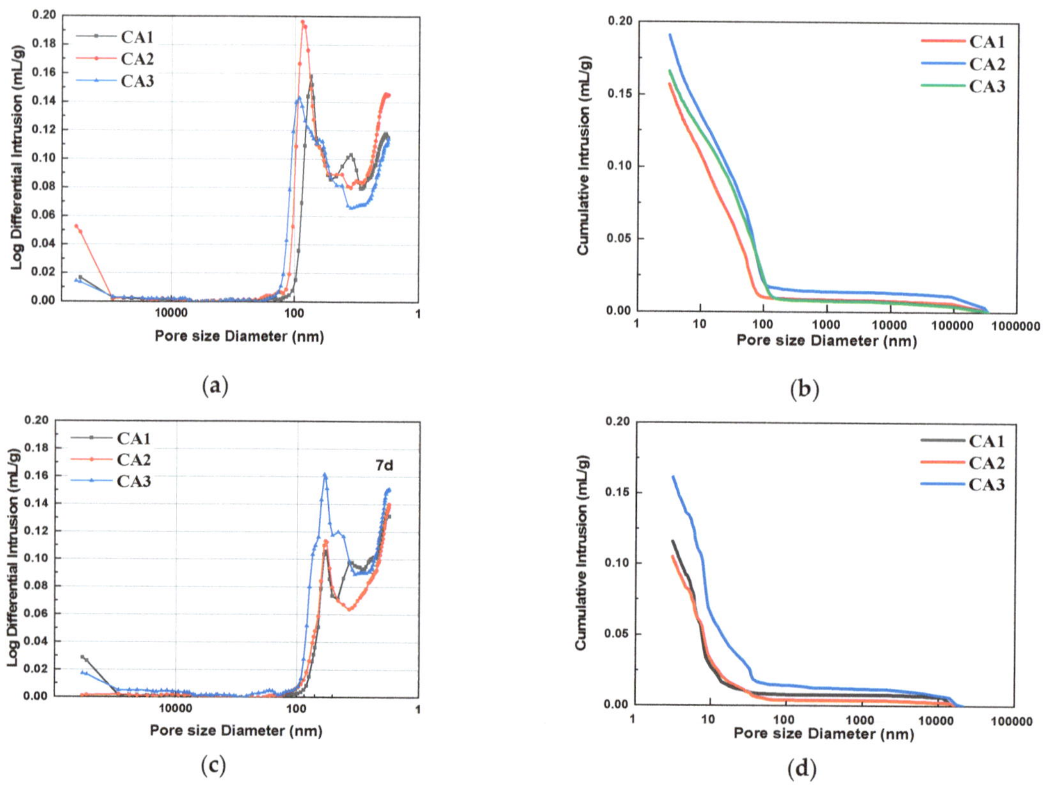

Figure 6. *Cont.*

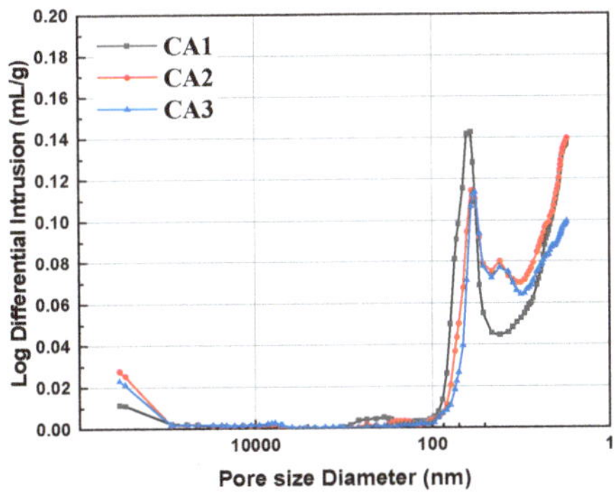

(e)

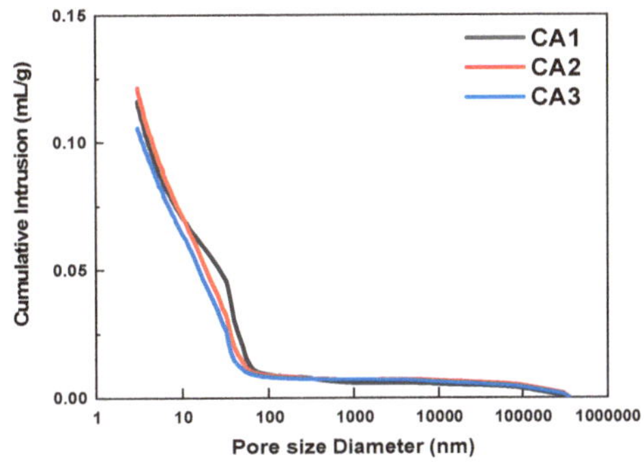

(f)

Figure 6. Differential pore size distribution and cumulative porosity of specimens with different CAMC contents at 3 d, 7 d, and 28 d. (**a**) Differential pore size distribution at 3 d; (**b**) cumulative porosity at 3 d; (**c**) differential pore size distribution at 7 d; (**d**) cumulative porosity at 7 d; (**e**) differential pore size distribution at 28 d; (**f**) cumulative porosity at 28 d.

3.6. SEM

Figure 7 shows the microstructure of samples with different CAMC dosages at different hydration ages. It can be observed from Figure 7 that at 3 d, more needle-like products, namely C-S-H, appeared in the hydration products of the CA1 sample, and the morphology distribution of the hydration products was more uniform compared to the CA3 sample, and the product structure was more dense. With the increase of CAMC dosage, the hydration effect of cement slurry is weakened, the generation of hydration products decreases, and the number of un-hydrated particles increases, covering the surface of the cement. The more CAMC is added, the more loose the product structure becomes. At the same time, the addition of CAMC significantly delays the generation of C-S-H. At 3 d, there are a large number of un-hydrated fly ash particles in all samples, and there are still many pores. The hydration product CH is less, and the structure is not dense enough. The hydration product of the CA1 sample contains more interlaced network C-S-H gel, the pores are relatively small, and the hydration product of clusters is the most. At 7 d, the introduction of CAMC resulted in the generation of many new clustered hydration products in the sample. Through analysis, it was found that this was due to the addition of CAMC, which transformed the single needle-like loose C-S-H grouting into a flocculent form. In pure water slurry, the direction of the C-S-H needle is usually disordered. The addition of CAMC can delay cement hydration, and the more the amount added, the better the delay effect, and the more the growth rate of needle-like C-S is inhibited. At 28 d, due to the change of nucleation and growth kinetics, C-S-H gel was wrapped by a large number of hydration products. In the later stage of hydration, the delayed hydration effect of CAMC almost disappears. At this time, the degree of cement hydration is high, the porosity is low, the structure is dense, and the microstructure of all samples is relatively consistent. Only a small number of large pores and scattered incomplete hydration of fly ash particles exist.

3.7. NMR

In order to further study the influence of CAMC content on the microstructure of hydration products of composite cementitious materials, the structure of C-S-H gel in the sample at 28 d hydration age was characterized by nuclear magnetic resonance (NMR).

Figure 8 shows the fitting results of ^{29}Si NMR peaks for samples with different CAMC contents at 28 d. It can be seen from Figure 8 that the sample doped with CAMC mainly contains three peaks: Q^0, Q^1, and Q^2. With the increase of CAMC content, the Q^2 peak strength increased significantly and the Q^0 unit peak strength decreased significantly, which indicated that a large amount of long-chain dimer C-S-H gel was formed in the later hydration period of the composite cementitious system, and the Q^0 peak strength decreased significantly, which indicated that the slurry had a high degree of hydration. Furthermore, in the CA3 sample, no Q1 was detected, indicating a low degree of hydration. The values of $I(Q^1)/I(Q^2)$ in all samples decrease with the increase of CAMC content, indicating that the degree of polymerization of silica tetrahedral structure gradually decreases, confirming that CAMC can delay cement hydration in the early stage of hydration.

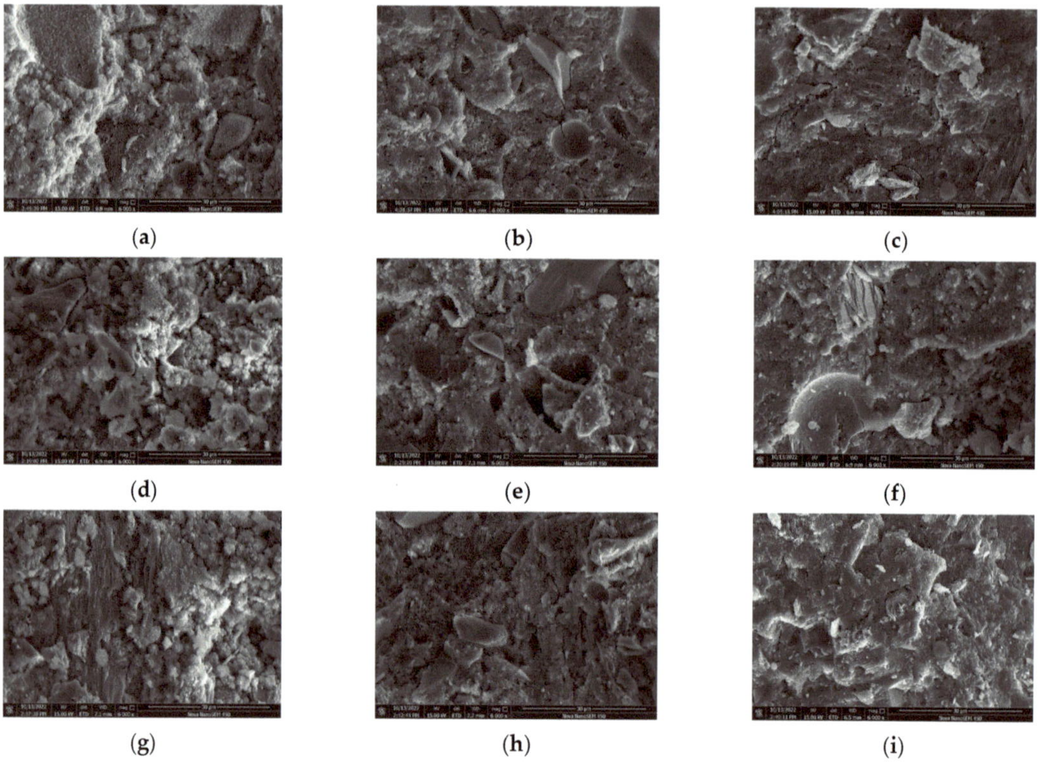

Figure 7. SEM of specimens with different CAMC contents at 3 d, 7 d, and 28 d. (**a**) CA1 at 3 d; (**b**) CA2 at 3 d; (**c**) CA3 at 3 d; (**d**) CA1 at 7 d; (**e**) CA2 at 7 d; (**f**) CA3 at 7 d; (**g**) CA1 at 28 d; (**h**) CA2 at 28 d; (**i**) CA3 at 28 d.

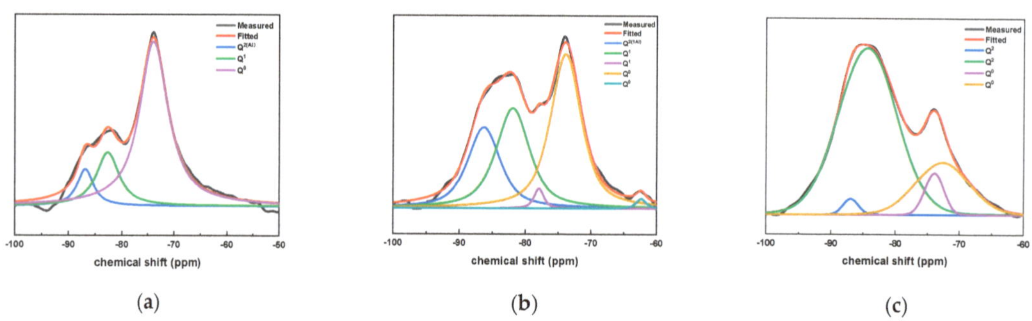

Figure 8. ^{29}Si NMR peak fitting results of specimens with different CAMC contents at 28 d. (**a**) CA1; (**b**) CA2; (**c**) CA3.

Furthermore, the peak area $I(Q^n)$ can be integrated according to the correlation peak intensity of the ^{29}Si NMR peak fitting of the sample, and the chemical structure of the C-S-H gel can be calculated accordingly. The calculation formula is shown in Table 5.

Table 5. C-S-H gel chemical structure parameter calculation formula [15,16].

Name	Equation
Degree of hydration	$\alpha = [1 - \frac{I(Q^0)}{\sum_{i=0}^{2} I(Q^2)}] \times 100\%$
Average chain length	$MCL = \frac{I(Q^1)+I(Q^2)+I[Q^2(1\text{Al})]}{1/2I(Q^1)+1/2I[Q^2(1\text{Al})]}$
Average Al/Si ratio	$\text{Al/Si} = \frac{1/2\text{I}[Q^2(1\text{Al})]}{I(Q^1)+I(Q^2)+I[Q^2(1\text{Al})]}$
Average silicon chain length	$ACL = \frac{I(Q^1)+I(Q^2)+3/2I[Q^2(1\text{Al})]}{1/2I(Q^1)}$

Note: Among them, $I(Q^0)$, $I(Q^1)$, $I(Q^2)$, and $I[Q^2(1\text{Al})]$ are the peak areas of Q^0, Q^1, Q^2, and Q^2(1Al), respectively.

According to the formula in Table 5, the hydration degrees of CA1, CA2, and CA3 samples at 28 d were calculated to be 73.5%, 76.1%, and 58.7%, respectively. This also confirms that CAMC can reduce the cement hydration rate, and within a certain range, the higher its content, the better the effect of delaying cement hydration.

The ^{27}Al NMR fitting results of the slurry with different CAMC dosages at 28 days of age are shown in Figure 9. It was found that the Al phase in all samples mainly consists of four coordinated and six coordinated structures. All samples have an obvious spectral peak near the chemical shift of 0 ppm. In addition, there is a clear peak near 6.6 ppm in each group, which is Al [6]-M in AFm, but there is no Al [6]-E in AFt, indicating that AFt has completely transformed into AFm. At the age of 28 days, each group of samples has an obvious formant at the chemical shift of 69 ppm, which corresponds to four coordinated aluminum. The Alenvironment is often similar to the low polymerized C-(A)-S-H gel, and the peak strength of Alis is also decreasing with the increase of CAMC content. This proves that the aluminum atom of Q^2 (1Al) exists in the form of Alcoordination in the sample, and the substitution rate of Al in the silica tetrahedron decreases with increasing CAMC content.

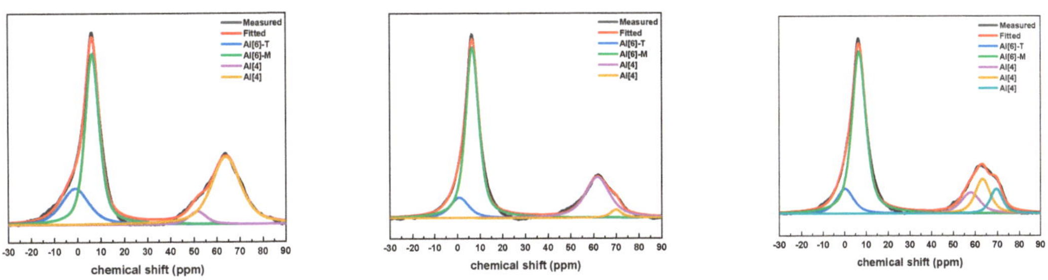

(a) (b) (c)

Figure 9. ^{27}Al NMR peak fitting results of specimens with different CAMC contents at 28 d. (**a**) CA1; (**b**) CA2; (**c**) CA3.

4. Conclusions

This article investigates the influence of low-temperature rise polymer CAMC on the hydration of composite cementitious material systems, revealing the regulatory mechanism of CAMC on them. Firstly, the effect of low-temperature rise polymers on the heat release during the hydration process of composite cementitious materials was studied through hydration heat. Then, based on the XRD data calculation of the hydration degree of cement clinker, a detailed analysis was conducted on the hydration product phase. Finally, the pore structure and microstructure of composite cementitious materials with different dosages of CAMC were characterized using micro testing methods such as MIP, SEM, and NMR. The conclusion is summarized as follows:

(1) The addition of CAMC can delay the hydration process of cementitious materials. With the increase of CAMC dosage, the first and second hydration exothermic peaks of the slurry are delayed and the peak intensity is significantly reduced. The first exothermic peak of the CA3 sample decreased by about 40% compared to the maximum exothermic rate of the CA1 sample, and the maximum exothermic peak appeared after a delay of about 13 h, indicating that CAMC can delay the hydration of cement and significantly reduce the hydration heat release rate of the slurry, and the higher the content within a certain range, the better the effect.

(2) CAMC has no effect on the type of hydration products but has a slight impact on the content of each phase of the hydration products. In the early stage of hydration, with the increase of CAMC content, the characteristic peak of CH appears delayed in time and its intensity decreases. At the age of 3 d, as the CAMC dosage increases from 0.1% to 0.3%, the CH peak intensity also gradually increases, while the C_2S and C_3S peak intensity decreases. In the XRD spectrum at 28 d, the peak positions of all samples are almost the same, and the intensities of CH, C_2S, C_3S, AFm, and other peaks are also equivalent.

(3) Based on the degree of hydration reaction of cement clinker and the changes in various components and CH and AFt contents, it can be seen that with the increase of CAMC dosage, the overall hydration rate of cement slows down and the degree of hydration reaction decreases. This effect is particularly evident in the early stages of hydration. The retarding effect of CAMC mainly acts on the silicon phase, which can reduce the CH content in the system in the early stage of hydration and delay cement hydration.

(4) As hydration progresses, the total porosity of each group of slurries decreases, and the volume of large pores decreases while the volume of small pores increases. In the early stage of hydration, the smaller the amount of CAMC added, the denser the structure of the hydration product. CAMC can make C-S-H gel form floccules, it can inhibit the formation of needle-like C-S-H in the early stage of hydration, and the delayed hydration effect of CAMC almost disappears in the late stage of hydration. As the content of CAMC increases, the pore size of the hardened slurry gradually refines, the proportion of large pores decreases, and the hydration microstructure becomes denser in the later stage.

(5) With the increase of CAMC dosage, the strength of the Q^2 peak in the slurry will decrease and the aluminum atom of Q^2 (^{1}Al) exists in the form of Alcoordination in the sample. CAMC can reduce its hydration degree and delay cement hydration.

Author Contributions: Conceptualization, L.W. and Z.Q.; methodology, J.W.; validation, G.S., H.W. and K.L.; formal analysis, X.D.; data curation, F.W.; funding acquisition, J.J. All authors have read and agreed to the published version of the manuscript.

Funding: The authors greatly acknowledge the National Science Fund for Distinguished Young Scholars, China (51925903), the General Program of the National Natural Science Foundation of China (52350004) the State Key Laboratory of High-Performance Civil Engineering Materials (2020CEM001) and the Science and Technology Research Project of China Railway (2020YY240610, K2020G033).

Data Availability Statement: The data presented in this study are available on request from the corresponding author. The data are not publicly available due to privacy.

Conflicts of Interest: Author Kai Liu was employed by the company Jiangsu China Construction Ready Mixed Concrete Co., Ltd. The remaining authors declare that the research was conducted in the absence of any commercial or financial relationships that could be construed as a potential conflict of interest.

References

1. Peschard, A.; Govin, A.; Grosseau, P.; Guilhot, B.; Guyonnet, R. Effect of polysaccharides on the hydration of cement paste at early ages. *Cem. Concr. Res.* **2004**, *34*, 2153–2158. [CrossRef]
2. Yang, Z.; Gao, Y.; Mu, S.; Chang, H.; Sun, W.; Jiang, J. Improving the chloride binding capacity of cement paste by adding nano-Al_2O_3. *Constr. Build. Mater.* **2019**, *195*, 415–422. [CrossRef]
3. Pourchez, J.; Ruot, B.; Debayle, J.; Pourchez, E.; Grosseau, P. Some aspects of cellulose ethers influence on water transport and porous structure of cement-based materials. *Cem. Concr. Res.* **2010**, *40*, 242–252. [CrossRef]
4. Patural, L.; Marchal, P.; Govin, A.; Grosseau, P.; Ruot, B.; Devès, O. Cellulose ethers influence on water retention and consistency in cement-based mortars. *Cem. Concr. Res.* **2011**, *41*, 46–55. [CrossRef]
5. Hou, P.; Cai, Y.; Cheng, X.; Zhang, X.; Zhou, Z.; Ye, Z.; Zhang, L.; Li, W.; Shah, S.P. Effects of the hydration reactivity of ultrafine magnesium oxide on cement-based materials. *Mag. Concr. Res.* **2017**, *69*, 1135–1145. [CrossRef]
6. Azenha, M.; Lameiras, R.; De Sousa, C.; Barros, J. Application of air cooled pipes for reduction of early age cracking risk in a massive RC wall. *Eng. Struct.* **2014**, *62–63*, 148–163. [CrossRef]
7. Ismail, M.; Noruzman, A.H.; Bhutta, M.A.R.; Yusuf, T.O.; Ogiri, I.H. Effect of vinyl acetate effluent in reducing heat of hydration of concrete. *KSCE J. Civ. Eng.* **2016**, *20*, 145–151. [CrossRef]
8. Khil, B.S.; Jang, S.J.; Choi, W.C.; Yun], H.D. Effect of barium-based phase change material (PCM) to control the heat of hydration on the mechanical properties of mass concrete. *Thermochim. Acta* **2015**, *613*, 100–107.
9. Won-Chang, C.; Bae-Soo, K.; Young-Seok, C.; Qi-Bo, L.; Hyun-Do, Y. Feasibility of Using Phase Change Materials to Control the Heat of Hydration in Massive Concrete Structures. *Sci. World J.* **2014**, *2014*, 781393.
10. Alkhraisat, M.H.; Rueda, C.; Jerez, L.B.; MariñO, F.T.; Torres, J.; Gbureck, U.; Cabarcos, E.L. Effect of silica gel on the cohesion, properties and biological performance of brushite cement. *Acta Biomater.* **2010**, *6*, 257–265. [CrossRef] [PubMed]
11. Pan, Z.H.; Cai, H.P.; Jiang, P.P.; Fan, Q.Y. Properties of a calcium phosphate cement synergistically reinforced by chitosan fiber and gelatin. *J. Polym. Res.* **2006**, *13*, 323–327. [CrossRef]
12. Wang, L.; Ju, S.; Wang, L.; Wang, F.; Sui, S.; Yang, Z.; Liu, Z.; Chu, H.; Jiang, J. Effect of citric acid-modified chitosan on the hydration and microstructure of Portland cement paste. *J. Sustain. Cem.-Based Mater.* **2023**, *12*, 83–96. [CrossRef]
13. Wang, L.; Wang, F.; Sui, S.; Ju, S.; Qin, Z.; Su, W.; Jiang, J. Adsorption capacity and mechanism of citric acid-modified chitosan on the cement particle surface. *J. Sustain. Cem.-Based Mater.* **2023**, *12*, 893–906. [CrossRef]
14. Wang, L.; Zhang, Y.; Guo, L.; Wang, F.; Ju, S.; Sui, S.; Liu, Z.; Chu, H.; Jiang, J. Effect of citric-acid-modified chitosan (CAMC) on hydration kinetics of tricalcium silicate (C_3S). *J. Mater. Res. Technol.* **2022**, *21*, 3604–3616. [CrossRef]
15. Richardson, I.; Groves, G. The structure of the calcium silicate hydrate phases present in hardened pastes of white Portland cement/blast-furnace slag blends. *J. Mater. Sci.* **1997**, *32*, 4793–4802. [CrossRef]
16. Andersen, M.D.; Jakobsen, H.J.; Skibsted, J. Characterization of white Portland cement hydration and the CSH structure in the presence of sodium aluminate by 27Al and 29Si MAS NMR spectroscopy. *Cem. Concr. Res.* **2004**, *34*, 857–868. [CrossRef]

buildings

MDPI

Article

Effects of Sodium Gluconate on the Fluidity and Setting Time of Phosphorus Gypsum-Based Self-Leveling

Xuepeng Shen [1], Hao Ding [1], Zhichun Chen [2], Ying Li [3], Wenxuan An [3], Aili Chen [4], Dongyi Lei [3,*], Ying Fang [1,*] and Dongxu Li [1,*]

[1] College of Materials Science and Engineering, Nanjing Tech University, Nanjing 210009, China; 202161103080@njtech.edu.cn (X.S.); 202161203249@njtech.edu.cn (H.D.)
[2] Technical Supervision and Research Center of the Building Materials Industry, Beijing 100024, China; czc910@163.com
[3] School of Civil Engineering, Qingdao University of Technology, Qingdao 266033, China; 13346092644@163.com (W.A.)
[4] Nanjing Kunchang New Material Co., Nanjing 210049, China; 15189837773@163.com
* Correspondence: leidongyi@qut.edu.cn (D.L.); powderfang@163.com (Y.F.); dongxuli@njtech.edu.cn (D.L.)

Abstract: To comprehensively utilize industrial by-products of gypsum while reducing the consumption of natural river sand, this experiment was conducted to prepare gypsum-based sandless self-leveling (PGSL) materials by using phosphorus-building gypsum (PBG) and portland cement (PC) as gelling raw materials with the addition of polycarboxylate superplasticizer (PCE), cellulose ethers (CE), and retarders. However, employing phosphogypsum as the source material results in a significant 30 min fluidity loss in the gypsum-based self-leveling system. Therefore, to enhance the flow characteristics of gypsum self-leveling, sodium gluconate was chosen for usage in this research. The impact of single and compound mixing of protein-based retarder (PR) and sodium gluconate (SG) on gypsum-based sandless self-leveling materials was evaluated in terms of heat of hydration analysis, pore structure, fluidity, strength, and setting time. According to the experimental findings, it was possible to considerably decrease the fluidity loss of gypsum-based sandless self-leveling materials, postpone the setting time, boost strength, and enhance pore structure when combined with 0.4% SG and 0.03% PR.

Keywords: gypsum self-leveling; fluidity loss; hydration; $CaSO_4$-$2H_2O$; phosphorus impurities

Citation: Shen, X.; Ding, H.; Chen, Z.; Li, Y.; An, W.; Chen, A.; Lei, D.; Fang, Y.; Li, D. Effects of Sodium Gluconate on the Fluidity and Setting Time of Phosphorus Gypsum-Based Self-Leveling. *Buildings* **2024**, *14*, 89. https://doi.org/10.3390/buildings14010089

Academic Editor: Xiaoyong Wang

Received: 3 December 2023
Revised: 22 December 2023
Accepted: 26 December 2023
Published: 28 December 2023

1. Introduction

Phosphogypsum is a solid waste obtained in the process of phosphorus fertilizer production, and its main component is $CaSO_4$-$2H_2O$. Subsequently, beta-hemihydrate gypsum (β-$CaSO_4$-$0.5H_2O$) and alpha-hemihydrate gypsum (α-$CaSO_4$-$0.5H_2O$) can be obtained using different calcination methods. According to the various qualities of phosphate ore, 1 ton can yield about 4–6 tons of phosphogypsum [1,2]. China is a significant producer of phosphogypsum production; the annual growth rate of phosphogypsum is about 70 million metric tons, of which the comprehensive utilization rate is only 15 percent. Due to the incomplete understanding of phosphogypsum in the early days, a lot of phosphogypsum was piled up near rivers and oceans, and the accumulation of a large amount of phosphogypsum caused significant damage to the surrounding environment, such as polluting the nearby soil and making the water source eutrophic, etc. [3–5]. Therefore, during the period of rapid national development, it is of great significance to seek new ways of complete utilization of phosphogypsum to improve the comprehensive utilization rate of phosphogypsum and to promote the construction of socialist modernization and civilization. Gypsum building materials prepared with phosphogypsum as the primary cementitious material can greatly reduce the amount of cement, reduce carbon dioxide emissions, are less harmful to the environment, and are a new type of green building material [6–8]. Among them, gypsum-based self-leveling mortar is a kind of dry powder material specially used for indoor leveling. It performs better in terms of acoustic isolation, thermal insulation, and moisture absorption, and provides micro-expansion after hydration to counteract the contraction produced by the hydration of cement components [9].

In traditional gypsum-based self-leveling materials, natural river sand is usually added as an aggregate filler. Still, due to the continuous development of society and urbanization, the consumption of wild river sand has significantly increased. The water–sand balance of the river has been altered by excessive sand mining, which has an impact on the river's stability and ecosystem and seriously threatens the safety of the river channel, the flood control levees, and the arable land [10–12], and most of the mechanism sand has an angular, rough surface, which, although it will improve the strength a little, is not conducive to mobility; we need to add too much water to improve mobility, which will again be unfavorable to the strength [13], and the actual construction of gypsum self-leveling is mostly done by machine pumping. The addition of sand will increase the gravity of self-leveling, which is not conducive to the efficiency of pumping. Consequently, it is imperative to prepare gypsum-based self-leveling materials without sand. In order to design gypsum-based self-leveling mortar, Zhi et al. [9] used high-strength gypsum to prepare gypsum self-leveling mortar and to research the impact of chemical admixtures on the mortar's setting time, liquidity, and mechanical qualities.

The first step in improving phosphogypsum utilization is to reduce the impact of phosphogypsum impurities on the gypsum. Among them, the biggest influences on the performance of phosphogypsum are phosphorus impurities and fluorine impurities. Among phosphorus impurities, phosphorus impurities mainly exist in the form of H_3PO_4, $Ca_3(PO_4)_2$, $FePO_4$-$2H_2O$, HPO_4^{2-}, $CaHPO_4^{-}2H_2O$, $H_2PO_4^-$, and $Ca(H_2PO_4)^{2-}H_2O$ [14,15]. Among them, soluble phosphorus has a more significant impact on the performance of gypsum, and the presence of soluble P_2O_5 significantly hinders the hydration, coagulation, and hardening processes of phosphorus gypsum and reduces the strength of its hardened body. In an acidic environment, soluble phosphorus mainly exists as H_3PO_4, $H_2PO_4^-$, and HPO_4^{2-}, while in an alkaline environment, phosphorus impurities mainly exist as PO_4^{3-}, which can easily combine with metal cations in the system to form inert insoluble phosphorus salts, which have much less effect on gypsum than soluble salts. Taken together, the magnitude of the impact on gypsum properties is $H_3PO_4 > H_2PO_4^- > HPO_4^{2-} > PO_4^{3-}$ [16]. At present, the gypsum-based self-leveling prepared with phosphogypsum is mostly β-$CaSO_4$-$0.5H_2O$, which has an irregular crystal structure compared with α-$CaSO_4$-$0.5H_2O$, and the gypsum-based self-leveling obtained from it has a lower strength and a larger loss of fluidity. Therefore, solving the excessive fluidity loss of phosphorus gypsum-based self-leveling has become a complex problem. Wang et al. [17] added calcium sulphoaluminate cement and three kinds of ground slag to gypsum self-leveling and found that two types of ground slag with small particle sizes can control the fluidity loss of 30 min to less than 3 mm. In addition to the slag and the cement in the appropriate ratio, it can also ensure a higher 1 d strength. Sliva et al. [18] added a highly effective polycarboxylate superplasticizer (PCE) to gypsum-based self-leveling at a suitable solubility and discovered that, despite the fact that the setting time does not adhere to the standard, the mechanical properties of gypsum self-leveling and the flow properties can be relatively good in the water–cement ratio of 0.5, glue–sand ratio of 1:0.5, and PCE of 1%.

Sodium gluconate is an organic compound with numerous industrial applications [19]. It can be used as a highly effective superplast and a highly effective retarder agent in the concrete industry, as well as an efficient chelating agent in construction, textile printing and dyeing, metal surface treatment, water treatment, cleaning of steel surfaces, and cleaning of glass bottles. Ma et al. [19] added SG to cement concrete and found that it increased the 3 d and 28 d compressive strengths, delayed setting time, and increased the fluidity of cement mortar. To enhance the performance of polycarboxylate-type (PCE) superplasticizers synergistically, Ren et al. [20] added SG to cement pastes. By grafting PCE side chains, they discovered that including SG improved PCE's fluidity and flow retention. A red-mud-based blended alkali-activated cement was given SG by Lin et al. [21] They discovered that it enhanced the fluidity of the paste and considerably slowed the pace of hydration. Numerous studies have indicated that SG greatly improves cement flow properties [19–21], but its reflection in gypsum is still relatively scarce. Typically, gypsum-based sandless self-leveling materials prepared from phosphogypsum usually have a significant loss of flow properties after 30 min. Gypsum self-leveling is usually pumped by machine in practical use, so the requirement for flow properties is very high. In this paper, SG and PR were used to improve this property of the material, and it was found that the addition of SG greatly improved the flowability after 30 min, improved the strength and other physical properties, evaluated the hydration process, porosity, and pore structure of the material, and finally successfully prepared a gypsum-based sandless self-leveling material with excellent properties.

2. Materials and Methods

2.1. Raw Materials

Guizhou Kailai Green Building Materials Co., Guizhou 551199, China, provided the phosphorus-containing building gypsum for this paper. It is grayish-white in appearance. Table 1 lists the fundamental physical characteristics, and Table 2 lists the chemical makeup.

Table 1. PBG's fundamental physical characteristics.

Water to Gypsum Ratio (%)	Setting Time (s)		Flexural Strength (MPa)		Compressive Strength (MPa)	
	Initial	Final	2 h	3 d-Dry	2 h	3 d-Dry
66	120	270	3.1	3.7	6.9	15.3

Table 2. The percentages of PBG and PC's chemical makeup.

Component	SO_3	CaO	SiO_2	P_2O_5	Al_2O_3	Fe_2O_3	Na_2O	K_2O	Loss
PBG	50.51	36.62	2.09	0.98	0.52	0.39	0.023	0.11	8.44
PC	2.31	57.67	20.69	0.41	6.3	5.74	0.1	0.52	9.93

The phase composition of PBG is depicted in Figure 1a, which demonstrates that $CaSO_4$-$0.5H_2O$ is the primary constituent of this PBG. According to GB/T 36141-2018 [22], the amount of $CaSO_4$-$0.5H_2O$ was calculated to be 87.57 wt%. Gypsum powder's apparent morphology as seen by scanning electron microscopy is depicted in Figure 1c, where it can be seen that the powder mostly takes the shape of a plate and contains a lot of irregularly shaped particles. Phosphorus, fluoride, alkali metal salt, and other organic contaminants were adsorbed onto the surface of the particles. The PBG particle size distribution is depicted in Figure 1b; the average particle size is 33.898 μm.

Anhui Conch Cement Co. manufactures this cement kind, which is PC 42.5. Purchased from Jiangsu Zhaojia Building Material Technology Co., Suzhou 215000, China, the superplasticizer is a polycarboxylate superplasticizer (PCE). Organosilicon defoamer (OD), a product of Jiangsu Zhaojia Building Materials Technology Co., Suzhou 215000, China, is an antifoam agent. Hydroxypropyl methylcellulose (CE), the ingredient in cellulose ether, can be purchased from Shenzhen Tongzhouda Co., Shenzhen 518000, China. Retarder was acquired from Jiangsu ZhaoJia Building Material Technology Co., Suzhou 215000, China, and is a polypeptide-bonded polymer (PR). SG (industrial grade) was purchased from Shandong Kaixiang Biotechnology Co., Shandong 262300, China.

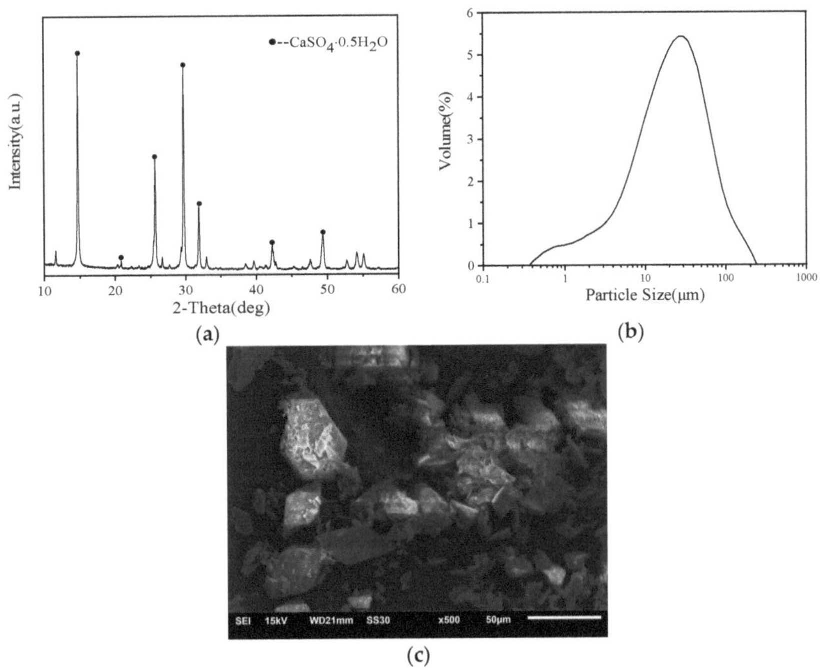

Figure 1. Characterization of PBG: (**a**) XRD; (**b**) particle size distribution; and (**c**) SEM image.

2.2. Sample Preparation Process

Table 3 lists the specific proportion design used in this paper: gypsum-based sandless self-leveling material admixture dosage based on the weight of dry powder, of which PCE is 0.15%, PR is 0.03%, OD is 0.1%, and CE is 0.03%, and the ratio of water to the amount of dry powder is 0.40.

Table 3. Mix proportions for the PGSL.

No.	PBG (wt%)	PC (wt%)	SG (wt%)	PR (wt%)
P0	95	5	0.04	0
K0	95	5	0	0.03
K1	95	5	0.01	0.03
K2	95	5	0.02	0.03
K3	95	5	0.03	0.03
K4	95	5	0.04	0.03
K5	95	5	0.05	0.03
K6	95	5	0.06	0.03

PBG was used as the primary cementitious material, and a small amount of PC was added to adjust the pH value of the system. The mass fraction of gypsum was 95%, and the mass fraction of cement was 5%. According to the test method stipulated in JC/T 1023-2021 [23], weighing was carried out by the proportions in Table 3, first stirring slowly for 60 s, scraping down the pan wall and mixing blade unevenly within 30 s, then stirring quickly for 60 s, and then stirring fast for 15 s after 60 s of static stopping until no bubbles were produced, to get the homogeneous slurry. Then, immediately loaded into a 40 mm by 40 mm by 160 mm cement sand test mold, the slurry is filled with a scraper to scrape flat. Then put the test mold into the constant temperature and humidity maintenance box (relative humidity $\geq$ 90%, temperature 20 °C $\pm$ 1 °C) for molding and maintenance. A flow chart for the specific preparation of gypsum self-leveling can be seen in Figure 2.

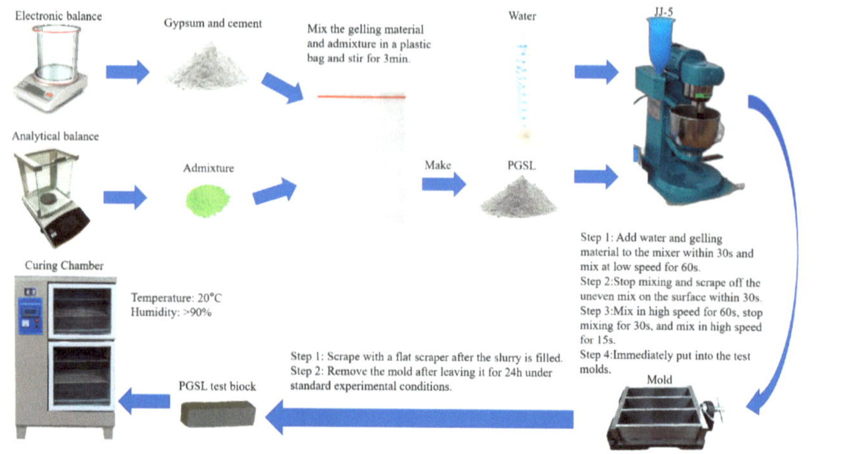

Figure 2. Flow chart for the preparation of gypsum-based self-leveling.

2.3. Test Method

2.3.1. Physical Properties

(1) Fluidity

According to the building materials industry standard JC/T 1023-2021 test, the test mold's fluidity was placed horizontally in the middle of the test plate, the slurry was stirred uniformly and poured into it, excess slurry was scraped off of the test mold's upper mouth, the test mold was raised vertically by 50–100 mm within 2 s, held for 10–15 s, and the slurry was allowed to flow freely for 4 min. Then the diameter of two vertical directions of the cake was measured, and the arithmetic average was taken to be accurate to 1 mm. The result was the initial fluidity. After 4 min of free flow of the slurry, measure the diameter of the two vertical directions of the cake, take the arithmetic mean value, accurate to 1 mm, and obtain the initial degree of flow. The above slurry was in the mixing bowl to stand for 30 min, mix for 30 s, and then poured into the two flowability test molds, respectively, to measure its flow; the 30 min fluidity was the arithmetic average of the above two fluidities.

(2) Setting time and strength

The sample's setting time was evaluated using the Chinese national standard GB/T 17669.4-1999 [24]. The sample's strength was evaluated in accordance with GB/T 17669.3-1999 [25]. The samples were dried at 40 °C in an oven until they reached a constant weight after curing for a predetermined amount of time. They were then loaded at a rate of 2.4 kN/s into an automatic compressive strength tester (Model AEC-201, Wuxi Alicon Instrument and Equipment Co., Wuxi 214000, China.) to test the samples' flexural and compressive strengths. The samples' flexural and compressive strengths were calculated.

(3) Rate of size change

The size change rate of the sample was tested according to the building materials industry standard JC/T 985-2005 [26]. After the sample was molded, it was placed under standard experimental conditions for (24 ± 0.5) h, the test direction was marked, and the length of the sample was tested in the marked direction within 30 min after demoulding, which was the initial length (L_0). Cured under standard experimental conditions for 28 d ± 8 h, the length of the specimen was tested according to the indicated direction, i.e., the length after natural drying (L_t). The rate of dimensional change of the sample was calculated using the following Equation (1)

$$\varepsilon = \frac{L_t - L_0}{L - L_d} \times 100\% \quad (1)$$

where ε is the rate of dimensional change. L_0 is the length of the specimen 24 h after moulding (mm), L_t is the length of the sample drying naturally for 28 d (mm), L is the length of the sample 160 mm, and L_d is the sum of the lengths of the two shrinkage heads buried in the slurry and the value of 20 ± 2 (mm). The average value of three specimens was taken as the final shrinkage rate.

2.3.2. Microscopic Properties

(1) Hydration process

A TAM AIR 8 calorimeter was used to measure the heat of the PGSL hydration process at 25 °C for 72 h, where the water–gel ratio was 0.4, in order to assess the exothermic behavior of the substance.

(2) Phase composition (XRD)

The physical structure of the samples was examined using an X-ray diffractometer (D/max-2200PC) from Rigaku, Japan, with a scanning rate of 10°/min and a scanning range of 10° to 80°.

(3) Pore structure analysis (MIP)

The pore structure of the PGSL was examined using a Poremaster GT-6.0 (Kantar Instruments USA, Inc., Lake Zurich, FL, USA), Quantum Chrome Mercuric Pressure instrument (MIP) from the United States after small sections of the center part had been removed and submerged in alcohol, dried, and evaluated. The measured pore size during the test ranged from 7.0 nm to 200.0 μm, while the incursion pressure ranged from 1.0 to 30,000.0 psi.

(4) Micromorphology (SEM)

After smashing the PGSL, the middle part was taken and glued to the conductive tape and gold plated. The samples were tested for microscopic properties using a scanning electron microscope (IT-300) from JEOL, Japan.

3. Results

3.1. *Effect of SG on the Setting Time of PGSL*

The impact of SG on the samples' setting time is depicted in Figure 3. It has been discovered that adding either SG or PR will lengthen the PGSL setting time. P0's beginning and final setting times are just 34 min and 43 min, which is considerably too little time and well below what is actually needed. In contrast, K0's initial and final setting times are 134 min and 216 min, respectively, which can accommodate the needs of actual use. Additionally, the initial setting time and final setting time of PGSL both significantly increase when PR and SG are combined, with the effect on the final setting time being more pronounced. This indicates that the time between the initial setting time and the final setting time is also growing. This is because PR itself has a large number of amino and carboxyl groups that chelate with Ca^{2+} in the slurry, on the one hand, reducing the Ca^{2+} solubility of the system; on the other hand, these chelates cover the surface of gypsum crystals, changing the way dihydrate gypsum ($CaSO_4$-$2H_2O$) crystals grow, thus delaying the nucleation process of the nucleus, which is then manifested as a prolongation of gypsum coagulation time [27]. Additionally, SG contains a variety of functional groups, like carboxyl and hydroxyl groups, that can react with Ca^{2+} in the system to form chelates. Because PR has a higher adsorption capacity than SG, its influence on retardation is more pronounced. When these two retarders are compounded, the adsorption capacity of the whole system on Ca^{2+} is greatly strengthened, and the macroscopic performance of PGSL is prolonged in the coagulation time.

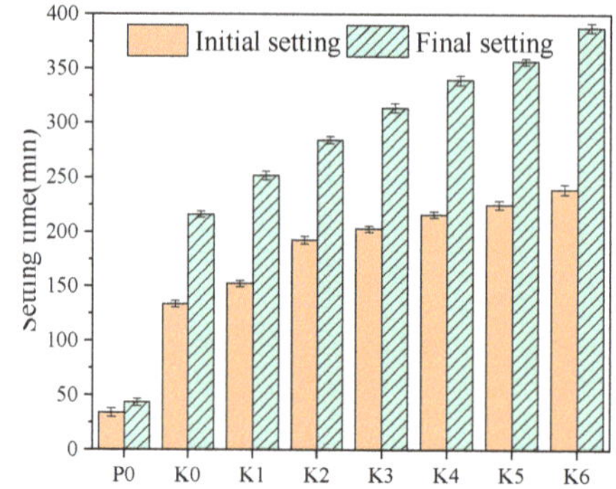

Figure 3. Effect of SG on the setting time of the samples.

3.2. *Effect of SG on the Fluidity of PGSL*

The impact of SG on the samples' fluidity is depicted in Figure 4. As depicted in Figure 4, the early fluidity of PGSL is specifically enhanced with an increase in SG dose. This is due to the addition of SG to a certain extent to promote the dispersion of the PCE [28], which assists in water reduction. So, when sodium gluconate is added, the initial fluidity of the self-leveling increases. In addition, the 30 min fluidity of P0 and K0 is almost lost. P0 is because if only SG is added, the retarding effect is relatively limited, and the 30 min is close to the initial setting time, so there is no fluidity after 30 min, whereas K0 is due to the impurities of the PBG itself, such as phosphoric acid, fluorine, and phosphoric acid impurities. The 30 min fluidity of PGSL is significantly improved after the compounding of PR and SG, which is increased to 120 mm at the addition of 0.01%. With the increase of SG, the fluidity increased, and the fluidity loss before and after 30 min is only 2 mm at the addition of 0.06%. Therefore, the addition of SG significantly improves the fluidity of PGSL.

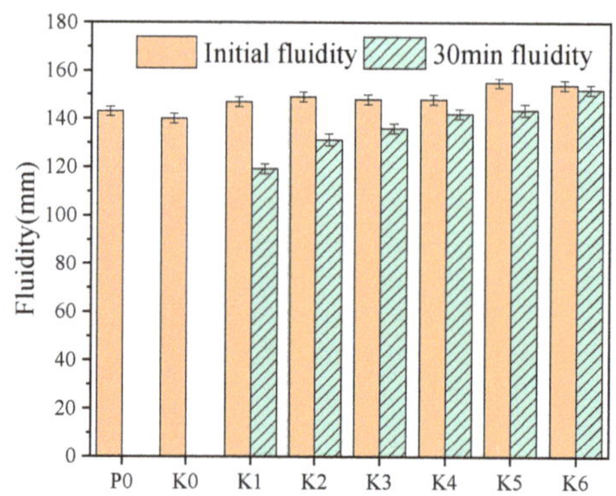

Figure 4. Effect of SG on the fluidity of the samples.

On the one hand, gypsum is acidic, while the addition of SG is weakly alkaline. After the addition of cement and the role of the system together to regulate the pH, gypsum in the alkaline environment, the H_3PO_4 transformed into H_2PO^{4-}, $HPO_4{}^{2-}$, and $PO_4{}^{3-}$, which have a much smaller impact on the performance of gypsum than H_3PO_4 [14–16]. However, many active functional groups, such as the carboxyl and hydroxyl groups in SG, react with free calcium ions to produce insoluble calcium salts, which precipitate on the surface of C_3A and $CaSO_4$-$0.5H_2O$ to form a calcium salt encapsulation layer. This effectively slows down the hydration rate of PGSL, in addition to SG incorporation at a level of less than 0.1%, which can promote the dispersion of PCE [28]. Under the combined effect of these three factors, the 30 min fluidity of PGSL is significantly improved by adding SG.

3.3. Effect of SG on the Mechanical Strength of PGSL

Figure 5 shows the effect of SG incorporation on the flexural and compressive strengths of PGSL. It can be seen that the late strength of K0 doped with PR alone is higher, while the early strength of P0 doped with SG alone is higher. The 1 d and 3 d flexural strengths of PGSL showed a trend of decreasing in the first rise and decreasing in the second rise with the increase of SG doping, while the 28 d strength showed a trend of increasing in the first rise and dropping in the second rise, with the flexural strength of K2 at 28 d, which is enhanced by 31.37%, showing the most significant increase. Gypsum self-leveling is combined with SG and PR, and it is discovered that the 28 d compressive strength of the self-leveling showed a rising and falling pattern with the rise of SG dosage, with the strength of K4 increasing most noticeably, with a 6.38% increase. The strength of K5 and K6 decreased more particularly, which may be due to the addition of excessive SG, which seriously slowed down gypsum's hydration degree and thus reduced the strength. Comprehensively, when the SG dosage reaches 0.04%, although it will reduce the early strength, it still meets the early strength requirements of JC/T 1023-2021, and it can improve a certain amount of late strength.

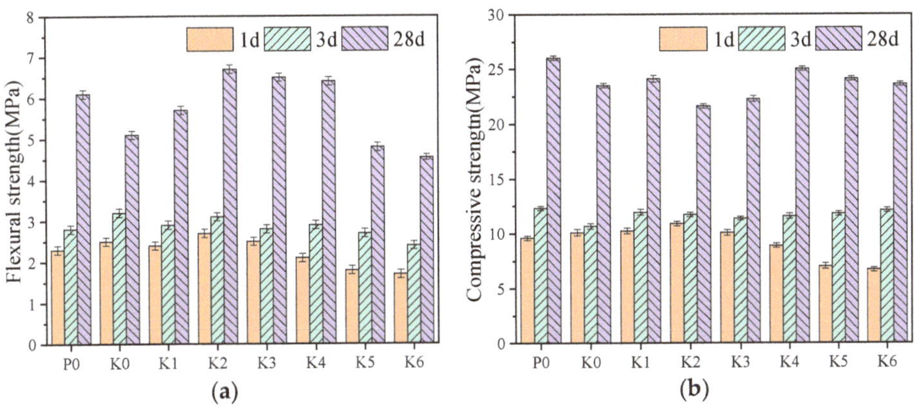

Figure 5. Effect of SG on the (**a**) Flexural strength and (**b**) Compressive strength of the samples.

3.4. Rate of Size Change at 28 d

Figure 6 shows the effect of SG incorporation on the rate of size change of PGSL. From the graph, it can be seen that the rate of dimensional change of PGSL shows a trend of decreasing and then increasing with the addition of SG, with the lowest rate of size change when the SG doping is 0.03% and 0.04%, i.e., K3 and K4. Since construction gypsum's theoretical water demand is 18.6% of its weight [29], the overall PGSL in this experiment demonstrated a phenomenon of shrinking. The amount of water given to PGSL in the experiment was significantly greater than the amount of water that was theoretically needed for it. As a result, during PGSL's subsequent hydration process, the excess water separated as water vapor, which caused PGSL to shrink. In this experiment, some PCs were added, which might lead to micro-expansion in the later hydration stage, but it was not enough to make up for the shrinkage caused by the gypsum. As a result, the PGSL as a whole exhibits a propensity for contraction. During subsequent hydration, this separates extra water from the PGSL as water vapor.

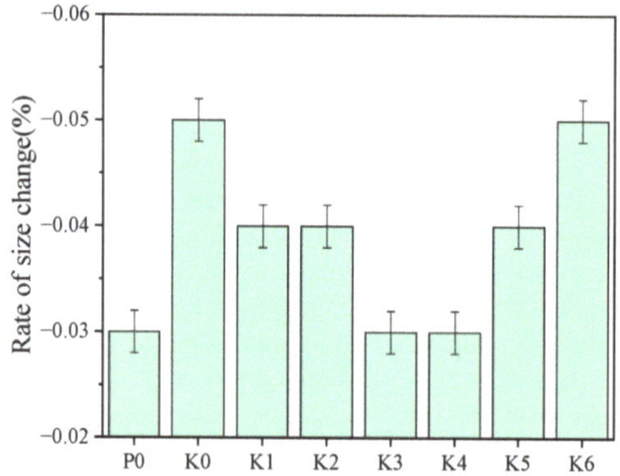

Figure 6. Effect of SG on the size change rate of the samples.

3.5. Heat of Hydration

The exothermic trend of hydration for various samples is shown in Figure 7. It is clear from the first peak in Figure 7a that the hydration of PBG occurs in stages. The $CaSO_4$-0.5H_2O is dissolved in water and comes into contact with it in the first step, releasing the heat of hydration. The rate of this stage is rather fast, and it is difficult to notice. Then, in the second stage, with the continual dissolution of $CaSO_4$-0.5H_2O in water and the accompanying precipitation of $CaSO_4$-2H_2O, the exothermic heat increases further. The stage of the accelerated phase is marked by a sharp temperature increase. The deceleration period represents the third stage, which occurs as the exothermic rate falls [30,31]. The heat of hydration diagrams show that the induction duration of gypsum lengthens with the addition of SG and that the peak of the heat of hydration of gypsum is delayed with an increase in the dose of SG. The addition of SG prolongs the coagulation time of PGSL. A second peak can be seen in the heat of the hydration curve as a result of the expansion of 5% PC in this experimental system. This is because PC has two exothermic peaks that correspond to the accelerated and decelerated phases of hydration, where the first peak coincides with the exothermic peak of gypsum. The inclusion of SG delays the second peak. The addition of SG hinders the hydration of C_3A and retards the hydration process of C_3A, which results in the delay of the second exothermic peak [19,32,33].

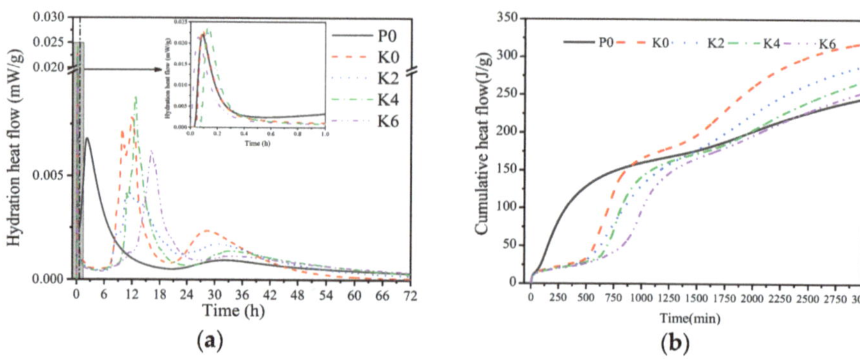

Figure 7. (**a**) Heat of hydration and (**b**) Cumulative heat release of the samples.

From Figure 6b, it can be seen that the early exothermic heat of hydration is less in K0 relative to P0, but the total exothermic heat is higher. The higher early cumulative exothermic heat in P0 may be because the single doping of SG does not significantly affect the hydration process of C_3A, which makes the early exothermic heat of PC and gypsum cumulative. Thus, the early cumulative exothermic heat in P0 is higher. Co-doping with

SG has a weakening effect on the cumulative exotherm of PGSL. According to general rules, the higher the heat of hydration, the greater the degree of hydration reaction, and the greater the fluidity loss [17,34]. However, as shown in Figure 6b, the heat of hydration decreases with an increase in SG, and the corresponding fluidity loss also decreases, which is consistent with Figure 4.

3.6. XRD

The XRD patterns of PGSL at 1 d, 3 d, and 28 d are displayed in Figure 8. The majority of the $CaSO_4$-$0.5H_2O$ has been hydrated into $CaSO_4$-$2H_2O$ after 1 d of hydration, as shown in Figure 7a. However, some of the $CaSO_4$-$0.5H_2O$ remains unhydrated and has noticeable diffraction peaks. As shown in Figure 6a, the diffraction peaks of $CaSO_4$-$2H_2O$ became more pronounced when SG was added, indicating that SG can postpone $CaSO_4$-$2H_2O$'s hydration. The XRD spectra of the PGSL hydrated for 3 d are displayed in Figure 7b. It is evident that at this point, $CaSO_4$-$2H_2O$ has completely hydrated to become $CaSO_4$-$2H_2O$, with the unique peak of $CaSO_4$-$2H_2O$ from K0 being more significant than that of P0. However, the $CaSO_4$-$2H_2O$ diffraction peaks are most effective at a 0.02% SG dosage when these two retarders are combined, while the distinctive peak is at its weakest at a 0.04% SG dosage. When the hydration process reaches 28 d, the diffraction peaks of $CaSO_4$-$2H_2O$ in K4 become most visible. This indicates that SG doping at 0.04% will postpone PGSL's early hydration but have minimal effect on its later hydration.

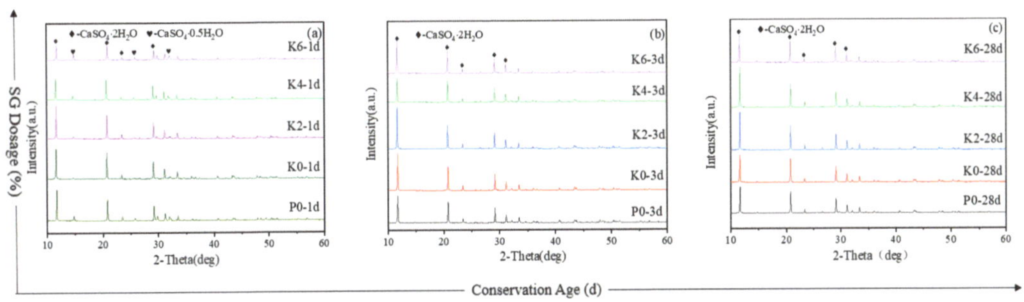

Figure 8. The XRD pattern of samples at (**a**) 1 d, (**b**) 3 d, and (**c**) 28 d.

3.7. Pore Structure

The physical properties of materials are closely related to their microscopic properties, and testing the pore structure is an essential means of testing the apparent microscopic properties. Generally speaking, the theoretical water requirement of $CaSO_4$-$0.5H_2O$ is only 18.6% of its weight. Still, in the production process of gypsum self-leveling, the actual amount of water added is much more considerable than its theoretical water requirement, which makes the excess water retained inside the gypsum slurry and precipitated as water vapor with the hydration process of gypsum, which leads to the emergence of many pores inside the gypsum [29]. Generally, pores with pore sizes >200 nm significantly influence gypsum properties [35]. The pore structure of PGSL is seen in Figure 9. Figure 9 shows that the SG-doped PGSL has significantly fewer detrimental pores than the PR-doped one. With the inclusion of SG during compounding, the detrimental pores and overall porosity inside the PGSL were greatly reduced, presumably because the gypsum hardened more quickly. In the early stages, the entire skeletal structure is produced, and as the gypsum continues to hydrate, it solidifies. The hydration of cement is greatly slowed down by the addition of SG, which has a negative impact on the gypsum's skeleton structure and increases porosity. In conclusion, adding SG can make PGSL's pore structure smaller.

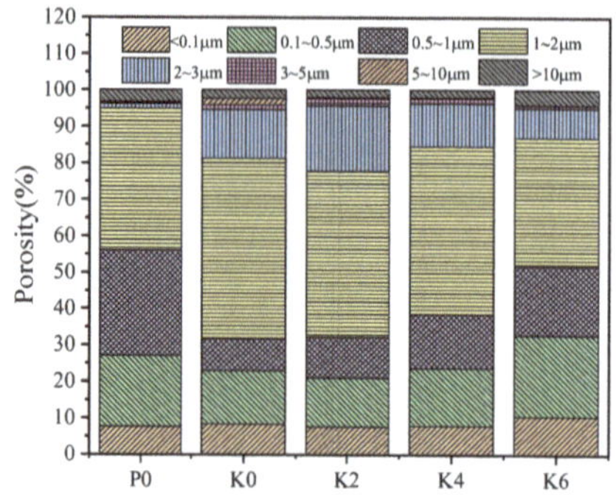

Figure 9. Pore size distribution of samples.

3.8. *Micromorphology*

The microscopic morphology of the PGSL samples after 1 d and 28 d of hydration is depicted in Figure 10. As seen in Figure 10a, the system's structure is still somewhat loose even after being hydrated to a level of 1d. These big plate-like crystals may represent $CaSO_4$-$0.5H_2O$ crystals that are still partially hydrated. It is evident from Figure 10b that the number of these partially hydrated $CaSO_4$-$0.5H_2O$ plate-like crystals increased following the addition of SG, confirming that the presence of SG delayed the hydration of $CaSO_4$-$0.5H_2O$ crystals. Figure 10c,d show that after the sample was hydrated for 28 d, the addition of SG caused the sample's porosity to decrease and the crystals to become more compact.

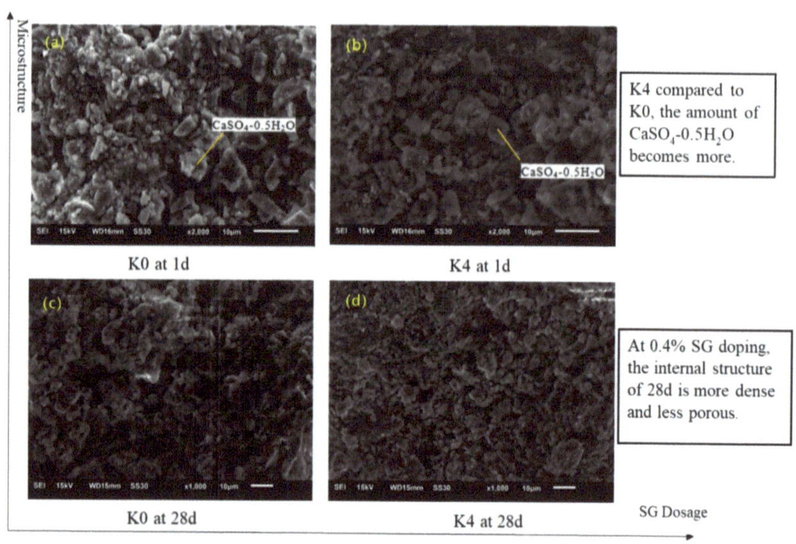

Figure 10. Effect of SG on the microscopic morphology of samples.

4. Discussion

This study examines the impact of SG and PR single and compound mixing on the gypsum's self-leveling qualities in an effort to enhance PGSL's operational performance. The following is a summary of the key findings:

(1) $CaSO_4$-$0.5H_2O$'s hydration can be postponed by SG and PR alike. When SG mixes with free calcium ions, it forms chelates that reduce the solubility of calcium ions in the solution and postpone the hydration of $CaSO_4$-$0.5H_2O$. The setting time of PGSL increases nearly linearly when the SG dosage is increased when SG and PR

are combined, and the difference between the initial and final setting durations gradually widens;

(2) The flow loss of both SG and PR is large; when these two admixtures are compounded, the working performance is significantly improved, and when the SG addition reaches 0.06%, the flow loss is the smallest, only 2 mm. This fully meets the needs of practical use;

(3) When 0.04% SG and 0.03% PR are compounded, the 1 d and 3 d strengths of gypsum self-leveling will decrease, but the 28 d strength will be improved, and the flexural and compressive strengths are improved by 25.49% and 6.38%, respectively;

(4) The addition of SG prolongs the induction period of $CaSO_4$-$0.5H_2O$ in PGSL and retards the hydration process of gypsum. In addition, it also significantly retards the hydration of C_3A, and the higher the dosage of SG, the more pronounced the retardation of hydration;

(5) When hydrated to 28 d, the total porosity of PGSL decreased, the average pore diameter became larger, and the microcrystalline structure was more dense when SG and PR were doped;

(6) The setting time was moderate, the mobility loss was 6 mm, the 1 d flexural and compressive strengths were 2.1 MPa and 8.9 MPa, and the 28 d flexural and compressive strengths were 6.4 MPa and 25 MPa when 0.04% SG and 0.03% PR were doped. The 28 d shrinkage rate was −0.03%, and all PGSL output during this period was of outstanding quality.

Author Contributions: Conceptualization, X.S.; methodology, X.S.; software, H.D.; validation, Z.C., W.A. and Y.L.; formal analysis, A.C.; investigation, Y.F.; resources, A.C.; data curation, Y.F.; writing—original draft preparation, X.S.; writing—review and editing, X.S. and D.L. (Dongyi Lei); supervision, Y.F.; project administration, D.L. (Dongyi Lei); funding acquisition, D.L. (Dongxu Li). All authors have read and agreed to the published version of the manuscript.

Funding: The National Natural Science Foundation of China (U23A20673, U22A20244), Qingdao West Coast New Area 2021 Science and Technology Plan Special Project (2021-102), Qingchuang Technology Project (2021KJ045), the Science and Technology Plan Projects of Sichuan Province (2021YFSY0002), and A Project Funded by the Priority Academic Program Development of Jiangsu Higher Education Institutions (PAPD) were jointly responsible for this work.

Data Availability Statement: Data are contained within the article.

Acknowledgments: The authors would like to thank Nanjing Kunchang new material Co. for providing the raw materials and experimental equipment.

Conflicts of Interest: Author Aili Chen was employed by the company Nanjing Kunchang New Material Co. The remaining authors declare that the research was conducted in the absence of any commercial or financial relationships that could be construed as a potential conflict of interest.

References

1. Chernysh, Y.; Yakhnenko, O.; Chubur, O.; Roubík, H. Phosphogypsum recycling: A review of environmental issues, current trends, and prospects. *Appl. Sci.* **2021**, *11*, 1575. [CrossRef]
2. Qin, X.; Cao, Y.; Guan, H.; Hu, Q.; Liu, Z.; Xu, J.; Hu, B.; Zhang, Z.; Luo, R. Resource utilization and development of phosphogypsum-based materials in civil engineering. *J. Clean. Prod.* **2023**, *387*, 135858. [CrossRef]
3. Gong, Y.; Dong, S.; Liu, L.; Wu, F. A sustainable composite cementitious material manufactured by phosphogypsum waste. *Appl. Sci.* **2022**, *12*, 12718. [CrossRef]
4. Calderón-Morales, B.R.S.; García-Martínez, A.; Pineda, P.; García-Tenório, R. Valorization of phosphogypsum in cement-based materials: Limits and potential in eco-efficient construction. *J. Build.* **2021**, *44*, 102506. [CrossRef]
5. Xue, S.; Li, M.; Jiang, J.; Millar, G.J.; Li, C.; Kong, X. Phosphogypsum stabilization of bauxite residue: Conversion of its alkaline characteristics. *J. Environ. Sci.* **2019**, *77*, 1–10. [CrossRef]
6. Chen, M.; Liu, P.; Kong, D.; Li, Y.; Chen, Y.; Cui, G.; Wang, J.; Yu, K.; Wu, N. Performance study and multi-index synergistic effect analysis of phosphogypsum-based composite cementitious material. *Coatings* **2022**, *12*, 1918. [CrossRef]
7. Li, Y.; Dai, S.; Zhang, Y.; Huang, J.; Su, Y.; Ma, B. Preparation and thermal insulation performance of cast-in-situ phosphogypsum wall. *J. Appl. Biomater.* **2018**, *16*, 81–92. [CrossRef]
8. Zhang, H.; Cheng, Y.; Yang, L.; Song, W. Modification of lime-fly ash-crushed stone with phosphogypsum for road base. *Adv. Civ. Eng.* **2020**, *2020*, 8820522. [CrossRef]

9. Zhi, Z.; Huang, J.; Guo, Y.; Lu, S.; Ma, B. Effect of chemical admixtures on setting time, fluidity and mechanical properties of phosphorus gypsum-based self-leveling mortar. *KSCE J. Civ. Eng.* **2017**, *21*, 1836–1843. [CrossRef]
10. Arulmoly, B.; Konthesingha, C.; Nanayakkara, A. Effects of blending manufactured sand and offshore sand on rheological, mechanicaland durability characterization of lime-cement masonry mortar. *Eur. J. Environ.* **2021**, *26*, 7400–7426. [CrossRef]
11. Hu, W. Impacts of human activities in the hanjiang river basin, China. *J. Coast. Res.* **2019**, *96*, 68–75. [CrossRef]
12. Padmalal, D.; Maya, K.; Sreebha, S.; Sreeja, R. Environmental effects of river sand mining: A case from the river catchments of Vembanad lake, southwest coast of India. *Environ. Geol.* **2007**, *54*, 879–889. [CrossRef]
13. Yang, X.; Li, C.; Wu, J. The application of blast furnace slag in cementitious self-leveling mortar. *Adv. Mat. Res.* **2011**, *328–330*, 1122–1126. [CrossRef]
14. Li, R.; He, W.; Duan, J.; Feng, S.; Zhu, Z.; Zhang, Y. Existing form and distribution of fluorine and phosphorus in phosphate rock acid-insoluble residue. *Environ. Sci. Pollut. Res.* **2022**, *29*, 7758–7771. [CrossRef] [PubMed]
15. Jiang, Y.; Kwon, K.D.; Wang, D.; Ren, C.; Li, W. Molecular speciation of phosphorus in phosphogypsum waste by solid-state nuclear magnetic resonance spectroscopy. *Sci. Total Environ.* **2019**, *696*, 133958. [CrossRef] [PubMed]
16. Zhang, J.; Wang, X.; Jin, B.; Liu, C.; Zhang, X.; Li, Z. Effect of soluble P_2O_5 form on the hydration and hardening of hemihydrate phosphogypsum. *Adv. Mater. Sci. Eng.* **2022**, *2022*, 1212649. [CrossRef]
17. Wang, Q.; Jia, R. A novel gypsum-based self-leveling mortar produced by phosphorus building gypsum. *Constr. Build. Mater.* **2019**, *226*, 11–20. [CrossRef]
18. Silva, D.B.P.; Lima, N.B.; Lima, V.M.E.; Estolano, A.M.L.; Nascimento, H.C.B.; Vilemen, P.; Padron-Hernández, E.; Carneiro, A.M.P.; Lima, N.B.D.; Povoas, Y.V. Producing a gypsum-based self-leveling mortar for subfloor modified by polycarboxylate admixture (PCE). *Constr. Build. Mater.* **2023**, *364*, 130007. [CrossRef]
19. Ma, S.; Li, W.; Zhang, S.; Ge, D.; Yu, J.; Shen, X. Influence of sodium gluconate on the performance and hydration of Portland cement. *Constr. Build. Mater.* **2015**, *91*, 138–144. [CrossRef]
20. Ren, J.; Fang, Y.; Ma, Q.; Tan, H.; Luo, S.; Liu, M.; Wang, X. Effect of storage condition on basic performance of polycarboxylate superplasticiser system incorporated sodium gluconate. *Constr. Build. Mater.* **2019**, *223*, 852–862. [CrossRef]
21. Lin, C.; Liu, Z.; Gao, Y.; Li, Z.; Zhang, J.; Niu, H. Study on the effect and mechanism of cement-based material retarder on red mud-based hybrid alkali activated cement. *J. Build.* **2023**, *70*, 106353. [CrossRef]
22. *GB/T 36141-2018*; Methods for Phase Composition Analysis of Calcined Gypsum. China Building Materials Industry Press: Beijing, China, 2018.
23. *JC/T 1023-2021*; Gypsum Based Self-Leveling Compound for Floor. China Building Materials Industry Press: Beijing, China, 2021.
24. *GB/T17669.4-1999*; Gypsum Pasters-Determination of Physical Properties of Pure Paste. China Building Materials Industry Press: Beijing, China, 1999.
25. *GB/T17669.3-1999*; Gypsum Pasters-Determination of Mechanical Properties. China Building Materials Industry Press: Beijing, China, 1999.
26. *JC/T 985-2005*; Cement-Based Self-Leveling Paddles for Floors. China Building Materials Academy: Beijing, China, 2005.
27. Ding, X.; Wei, B.; Deng, M.; Chen, H.; Shan, Z. Effect of protein peptides with different molecular weights on the setting and hydration process of gypsum. *Constr. Build. Mater.* **2022**, *318*, 126185. [CrossRef]
28. Tan, H.; Zou, F.; Ma, B.; Guo, Y.; Li, X.; Mei, J. Effect of competitive adsorption between sodium gluconate and polycarboxylate superplasticizer on the rheology of cement paste. *Constr. Build. Mater.* **2017**, *144*, 338–346. [CrossRef]
29. Jiao, J.; Shen, S.; Ding, H.; Lu, D.; Li, D. Study on the hydration and properties of multiphase phosphogypsum synergistically activated by sodium sulfate and calcium sulfate whisker. *Constr. Build. Mater.* **2022**, *355*, 129225. [CrossRef]
30. Hu, K.; Sun, Z. Influence of polycarboxylate superplasticizers with different functional units on the early hydration of C_3A-Gypsum. *Materials* **2019**, *12*, 1132. [CrossRef] [PubMed]
31. Tydlitát, V.; Tesárek, P.; Černý, R. Effects of the type of calorimeter and the use of plasticizers and hydrophobizers on the measured hydration heat development of FGD gypsum. *J. Therm. Anal.* **2008**, *91*, 791–796. [CrossRef]
32. Florian, A.H.; Johann, P. Impact of aging on the hydration of tricalcium aluminate(C_3A)/gypsum blends and the effectiveness of retarding admixtures. *Z. Naturforsch. B* **2020**, *75*, 739–753. [CrossRef]
33. Zhang, X.; He, Y.; Lu, C.; Huang, Z. Effects of sodium gluconate on early hydration and mortar performance of portland cement-calcium aluminate cement-anhydrite binder. *Constr. Build. Mater.* **2017**, *157*, 1065–1073. [CrossRef]
34. Zhao, H.; Sun, W.; Wu, X.; Gao, B. The properties of the self-compacting concrete with fly ash and ground granulated blast furnace slag mineral admixtures. *J. Clean. Prod.* **2015**, *95*, 66–74. [CrossRef]
35. Zhang, X.; Zhou, S.; Zhou, H.; Li, D. The effect of the modification of graphene oxide with gamma-aminopropyltriethoxysilane (KH550) on the properties and hydration of cement. *Constr. Build. Mater.* **2022**, *322*, 126497. [CrossRef]

buildings

MDPI

Review

Feasibility of Preparing Steel Slag–Ground Granulated Blast Furnace Slag Cementitious Materials: Synergistic Hydration, Fresh, and Hardened Properties

Jianwei Sun [1,2,*], Shaoyun Hou [1,2], Yuehao Guo [1,2], Xinying Cao [1,2] and Dongdong Zhang [3]

[1] School of Civil Engineering, Qingdao University of Technology, Qingdao 266520, China; houshaoyun19980120@163.com (S.H.); kiki2438199144@163.com (Y.G.); lc211008@126.com (X.C.)
[2] Engineering Research Center of Concrete Technology under Marine Environment, Ministry of Education, Qingdao 266520, China
[3] Jinan Rail Transit Group Co., Ltd., Jinan 250000, China; 15628861197@163.com
* Correspondence: jianwei_68@126.com

Abstract: Steel slag and GBFS are wastes generated during the steel and iron smelting process, characterized by their considerable production rates and extensive storage capacities. After grinding, they are often used as supplementary cementitious materials. However, the intrinsic slow hydration kinetics of steel slag–GBFS cementitious material (SGM) when exposed to a pure water environment result in prolonged setting times and diminished early-age strength development. The incorporation of modifiers such as gypsum, clinker, or alkaline activators can effectively improve the various properties of SGM. This comprehensive review delves into existing research on the utilization of SGM, examining their hydration mechanisms, workability, setting time, mechanical strengths, durability, and shrinkage. Critical parameters including the performance of base materials (water-to-cement ratio, fineness, and composition) and modifiers (type, alkali content, and dosage) are scrutinized to understand their effects on the final properties of the cementitious materials. The improvement mechanisms of various modifiers on properties are discussed. This promotes resource utilization of industrial solid wastes and provides theoretical support for the engineering application of SGM.

Keywords: steel slag; GBFS; hydration; mechanical property; durability; shrinkage

Citation: Sun, J.; Hou, S.; Guo, Y.; Cao, X.; Zhang, D. Feasibility of Preparing Steel Slag–Ground Granulated Blast Furnace Slag Cementitious Materials: Synergistic Hydration, Fresh, and Hardened Properties. *Buildings* **2024**, *14*, 614. https://doi.org/10.3390/buildings14030614

Academic Editor: Mizan Ahmed

Received: 5 January 2024
Revised: 5 February 2024
Accepted: 7 February 2024
Published: 26 February 2024

1. Introduction

Steel slag, a nonmetallic by-product generated during the steelmaking process, accounts for approximately 15–20% of crude steel production [1–3]. Globally, steel slag production is estimated to be around 190–280 million tons per year, with China contributing approximately 50% of the total and its production continuing to rise [4–6]. With proper utilization, steel slag can not only contribute to resource conservation but also reduce environmental pollution. Additionally, the comprehensive utilization of steel slag has the potential to lower production costs and enhance economic benefits. The utilization of steel slag in Japan, Europe, the United States, and China is shown in Figure 1. Steel slag can be disposed of in road engineering, cement production, internal recycling, civil engineering, and agriculture in Japan, Europe, the United States, and China [2–6]. It is noteworthy that 30–50% of steel slag is used for road construction in other countries, while more than 70% of steel slag is for final disposal. Therefore, compared to developed countries such as the United States and Japan, which achieve nearly 100% utilization rates, the actual comprehensive utilization rate of steel slag in China is only 30% [2,3]. The low utilization rate of steel slag results in a significant accumulation of waste, occupying land resources and posing environmental risks because of the leaching toxicity of heavy metals like Mn and Cr. This situation impedes the sustainable development of the steel industry [7–9]. Consequently, it has become crucial to address how to increase the comprehensive utilization rate of steel slag and transform it from waste into a valuable resource.

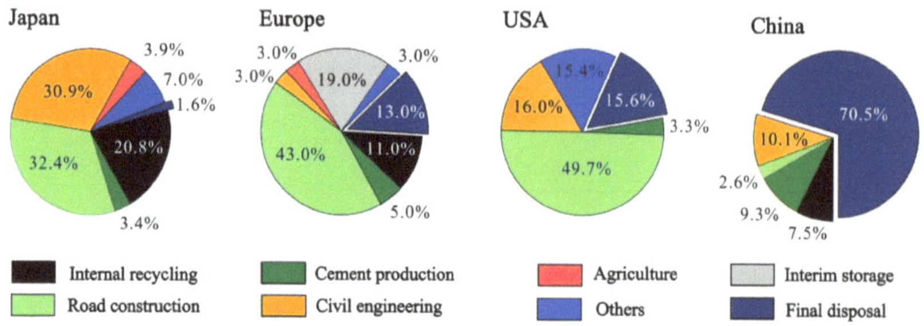

Figure 1. Steel slag utilization in Japan, Europe, the United States, and China [2].

The predominant form of steel slag in China is derived from the converter process, composing approximately 70% of the accumulated steel slag reserves. Steel slag contains a large amount of inactive components, mainly composed of the RO phase (MgO-FeO-MnO solid solution) [7–13]. Its active components, including tricalcium silicate (C_3S), dicalcium silicate (C_2S), calcium aluminate, and calcium iron aluminate, are similar to the active components of Portland cement clinker [7–13]. Owing to these similarities, steel slag exhibits a measure of hydraulicity, which has led to its designation as a 'substandard cement' or low-quality cement. This identifies it as a prospective sustainable cementitious material with inherent green qualities. However, because of the limited number of active phases in steel slag, when it is used solely as a cementitious material, it has inherent drawbacks such as a slow reaction rate, low degree of reaction, and slow development of early strength [14,15]. Given the shared metallurgical lineage between steel slag and ground granulated blast furnace slag (GBFS), the former is being considered for cotreatment with the latter to improve its performance as a cementitious material. The main phase of GBFS is amorphous and has high reactivity. When steel slag is mixed with GBFS, the clinker-like characteristics of steel slag and the pozzolanic reaction characteristics of GBFS can mutually promote the hydration process, significantly improving the hydration degree of the composite system, especially the later strength of the composite material. However, the application of steel slag–GBFS cementitious material (SGM) remains limited due to extended setting times and inadequate early strength, which fail to satisfy the demands of various engineering requirements [16–28]. A bibliometric study based on the Web of Science database is finished and shown in Figure 2. As can be seen, there has been an increasing amount of research on SGM in the past 15 years. In particular, after 2020, the number of published papers doubled compared to before. This indicates that scholars are increasingly valuing the development of SGM and have high expectations for this low-carbon material. To remedy these deficiencies of SGM, modifiers such as gypsum, cement clinker, and alkaline activators are applied. From statistical data, it can be seen that with the development of alkali-activated cementitious materials, over 50% of studies have used alkaline activators as effective modifiers of SGM. The proportion of research on various types of gypsum as modifiers is 21.05%. However, the research on cement clinker as a modifier is only 9.47%. Using modifiers accelerates the development and application of SGM in the field of cementitious materials. From the perspective of green development and solid waste utilization, further research on using by-product gypsum as a modifier is needed.

Based on extensive research by both domestic and international scholars on SGM, this paper analyzes and summarizes the existing research results from the perspectives of hydration characteristics, setting time, and workability of fresh mortar, mechanical properties, durability, and volume shrinkage of hardened mortar. Furthermore, the paper delves into the effects of modifiers on the performance of SGM. It elucidates how these materials improve the overall characteristics of cementitious materials while also providing an in-depth analysis of the underlying mechanisms that contribute to this enhancement. Finally, the existing issues in current research and future development trends are analyzed and discussed, aiming to provide theoretical support for the application and promotion of SGM.

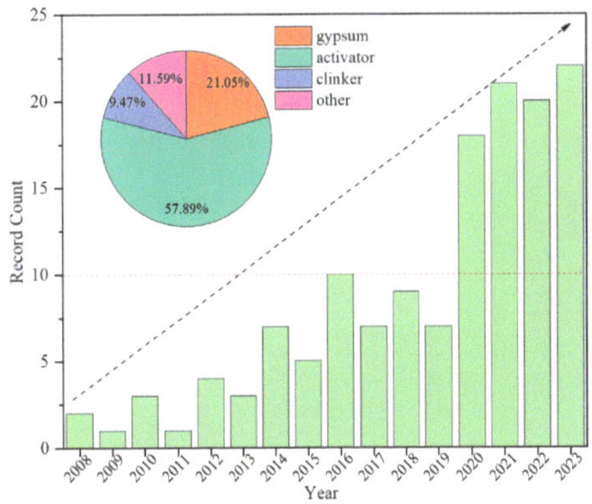

Figure 2. Record count in the field of SGM.

2. Synergistic Hydration of SGM

Steel slag itself exhibits a certain degree of hydraulicity and can react with water at room temperature, while GBFS has limited reactivity with water and requires activation in an alkaline environment with a pH value higher than 12.5 [29–32]. Studies have found that when steel slag and GBFS are mixed, there is a synergistic effect in terms of hydration [33,34]. The hydration mechanism of SGM is illustrated in Figure 3. Upon contact with water, active components such as calcium aluminate and C_3S in steel slag dissolve, releasing Ca^{2+}, OH^-, and small amounts of Al^{3+} and Si^{4+}. This results in an increase in the pH of the pore solution, represented by the "steep and high" first exothermic peak (S1) in the heat evolution rate curve [34]. After a short induction period (S2), the early hydration of the small amount of calcium aluminate, C_3S, and β-C_2S in steel slag leads to the acceleration of hydration and the formation of the second peak (S3) [34]. At this stage, Si-O-Al and Si-O-Si bonds in the GBFS are broken in the alkaline environment [34]. The hydration products of steel slag, namely $Ca(OH)_2$ and C-S-H gel, precipitate and cover the surfaces of C_3S and C_2S particles, thereby reducing the hydration rate and entering the deceleration period (S4) [34]. After a long stabilization period (S5), the hydration of Si-O-Al and Si-O-Si bonds in the GBFS, along with the reaction between $[SiO_4]^{4-}$ and $[AlO_4]^{5-}$ released from GBFS and $Ca(OH)_2$ in the system, lead to the reacceleration of the reaction rate and the appearance of the third exothermic peak (S6) [34]. As hydration progresses, two types of gel products, namely calcium silicate hydrate (C-S-H) gel and calcium aluminate hydrate (C-A-H) gel, accumulate on the surfaces of mineral particles, and the rate of volcanic ash reaction gradually decreases and tends to stabilize (S7) [34–40]. Some researchers have suggested that the highest reaction degree is achieved when the mass ratio of steel slag to GBFS is 1:1 or 2:3 [16–20]. It is important to note that the products of volcanic ash reaction are not solely C-S-H and C-A-H gels but also include calcium aluminosilicate hydrate (C-A-S-H) gel.

Due to the limited alkalinity provided by the self-hydration of steel slag, SGM is known to have slow reaction rates and a lack of early strength. In order to address this issue, SGM is commonly modified with materials such as gypsum (natural gypsum and industrial by-product gypsum), cement clinker, or alkaline activators (e.g., sodium silicate, $Ca(OH)_2$, NaOH, and Na_2CO_3) [41–49]. $Ca(OH)_2$, formed by the hydration of cement clinker and OH^- provided directly by alkaline activators, creates a more alkaline environment for cementitious materials, promoting the dissociation of $[SiO_4]^{4-}$ and $[AlO_4]^{5-}$ in the GBFS and further improving the early strength of materials. However, excessive amounts of certain modifiers, such as excessive cement clinker, NaOH, and Na_2CO_3, may actually reduce the later strength of materials. This is because in the later stages of hydration, excessive OH^- or CO_3^{2-} combine with Ca^{2+} on the surface of particles to form excessive $Ca(OH)_2$ and $CaCO_3$, hindering continuous hydration of both steel slag and GBFS particles [42]. On the other hand, sodium silicate plays a dual activation role by providing OH^- and $[SiO_4]^{4-}$, thereby not only offering an alkaline environment but also supplying the necessary silicon source for reactions, ensuring the development of later strength [50–52].

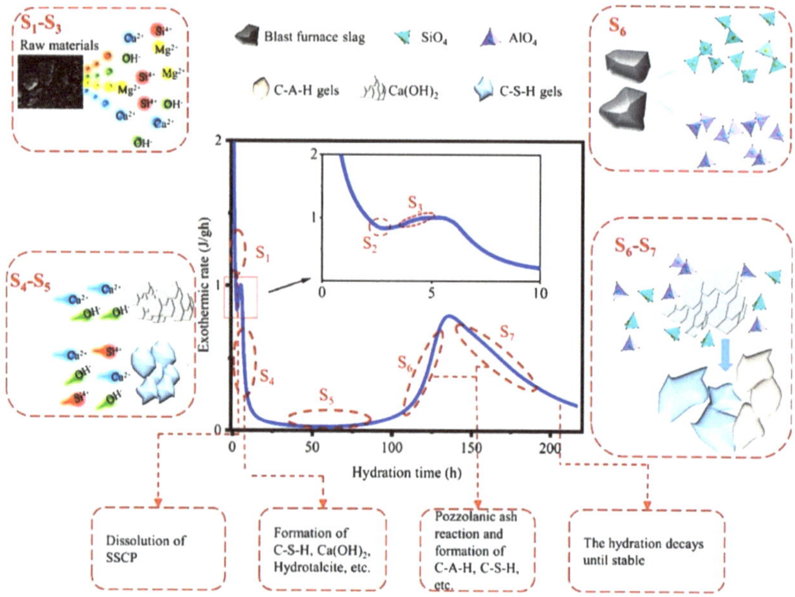

Figure 3. Hydration mechanism of SGM [34].

Gypsum is an effective modifier, and steel SGM with gypsum system exhibits excellent performance. Studies have shown that steel slag, GBFS, and gypsum exhibit significant synergistic effects during the hydration process [53,54]. These effects are mainly attributed to the reactions where the OH^- provided by steel slag, the Al^{3+} from steel slag and GBFS, and the SO_4^{2-} from gypsum react with the Ca^{2+}. This reaction causes the formation of insoluble calcium aluminate (AFt), driving the continuous progress of the reaction [55]. Xu et al. [56] investigated the effects of three types of industrial by-product gypsum, namely desulfurization gypsum (DG), desulfurization ash (DA), and fluorogypsum (FG), on the hydration and hardening properties of SGM. The results indicate that in the early stages, needle-rod-shaped AFt and amorphous gel are formed in all three cementitious systems. However, the DA system, which has a lower gypsum content, shows a significantly lower amount of AFt compared to the other systems. In the later stages, both the DG system and the FG system exhibit similar structures, characterized by interlaced AFt and gel filling and encapsulation. In contrast, the DA system only shows gel formation without a distinct AFt structure. As a result, the presence of AFt in the DG and FG systems contributed to the higher early strength of SGM. The microstructures of the hardened pastes in the DG and DA systems are shown in Figure 4. The gel morphology differs significantly between the two systems. The gel in the DG system appears cluster-like (Figure 4a), while the gel in the DA system exhibits a network-like structure due to the relatively higher amount of space provided by the lower AFt content (Figure 4b). The hydration mechanism of these three composite systems using the DG system as an example is shown in Figure 5. In the initial stages, the hydration of steel slag (Equation (1)) and the dissolution of DG release OH^-, Ca^{2+}, and SO_4^{2-}. At the same time, under alkaline conditions, Si-O-Al and Si-O-Si bonds in the GBFS are broken, resulting in the formation of $[SiO_4]^{4-}$ and $[AlO_4]^{5-}$

monomers. Subsequently, the reaction between Ca^{2+}, $SO_4{}^{2-}$, and $[AlO_4]^{5-}$ produces AFt (Equation (2)), while the reaction between Ca^{2+} and $[SiO_4]^{4-}$ leads to the formation of C-S-H gel (Equation (3)). Finally, the continuous production of C-S-H gel fills and encapsulates the interwoven AFt, ensuring the development of later strength. Therefore, the ultimate products of SGM with gypsum are C-S-H gel and AFt [56].

$$3CaO{\cdot}SiO_2 + SiO_2 + xH_2O \rightarrow xCaO{\cdot}SiO_2{\cdot}yH_2O + (3 - x)Ca(OH)_2 \tag{1}$$

$$12Ca^{2+} + 6SO_4{}^{2-} + 2AlO_4{}^{2-} + 2OH^- + 61H_2O \rightarrow 2(3CaO{\cdot}Al_2O_3{\cdot}3CaSO_4{\cdot}32H_2O) \tag{2}$$

$$aCa^{2+} + SiO_4{}^{4-} + bH_2O \rightarrow aCaO{\cdot}SiO_2{\cdot}bH_2O \tag{3}$$

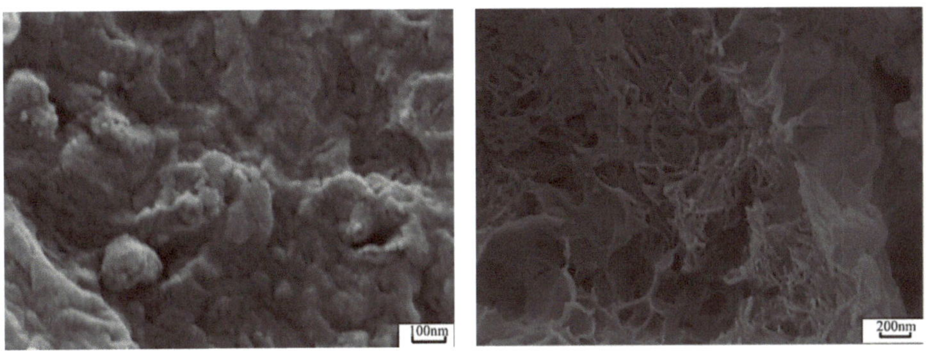

(**a**) (**b**)

Figure 4. Microstructures of steel slag–GBFS-IBG hardened pastes: (**a**) gel in DG; (**b**) gel in DA [56].

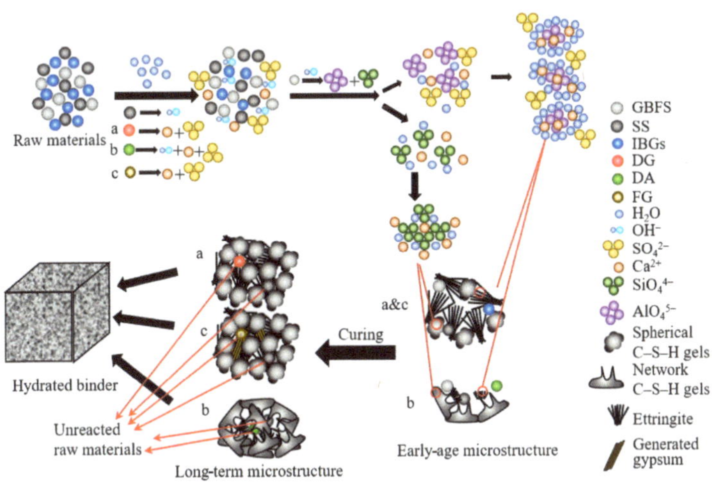

Figure 5. Hydration mechanism of steel slag–GBFS-IBG cementitious material [56].

When cement clinker is used as a modifier, its effectiveness is limited. Zhao et al. [57] prepared various SGMs and compared them with the Portland cement system to study the differences in hydration processes and pore structures. Overall, the cumulative heat of SGM at all ages is lower than that of Portland cement, and the cumulative heat within 72 h is less than 50% of that observed in the Portland cement system. This indicates that the alkaline environment provided solely by the hydration of clinker is relatively weak and insufficient to significantly enhance the initial hydration rate and reaction degree of steel slag and GBFS [58]. Regarding the pore structure, the total porosity of hardened SGM pastes is observed to be higher compared to that of hardened Portland cement pastes. Thus, the ternary system exhibits lower density and significantly lower strength when compared to the Portland cement system. These findings suggest that the use of cement clinker as a modifier in SGM may have limitations in terms of improving the hydration rate and enhancing the mechanical properties of materials. Therefore, alternative modifications such as the addition of gypsum or alkaline activators should be considered to optimize the performance of SGM.

When alkaline activators are used as modifiers, they generally provide a higher alkaline environment than cement clinker, resulting in a higher hydration degree of steel slag and GBFS. You et al. [47] studied the hydration process of alkali-activated SGM and compared the hydration rate and cumulative heat with the Portland cement system (Figure 6). The heat rate and cumulative heat of alkali-activated GBFS materials exhibit similar hydration processes to those of Portland cement. However, alkali-activated SGM has a lower and delayed peak value of the main peak compared to the alkali-activated GBFS material. This is primarily due to the lower reactivity of steel slag, which causes a delay in the appearance and reduces the peak value of the main exothermic peak. The cumulative heat within 7 h of the alkali-activated SGM is higher than that of the Portland cement system, but within 72 h, it was only 53% of that of Portland cement. This indicates that alkali-activated SGM has a higher early-stage heat and functions as a low-heat cementitious material. Furthermore, the incorporation of retarders effectively delays the occurrence of the main exothermic peak and reduces the hydration rate of alkali-activated SGM, prolonging the setting time of materials. In terms of hydration products, C-A-S-H gel is the main product in alkali-activated SGM. The relative content of steel slag and GBFS does not change the type of hydration products but can affect the structure of gel products. With an increase in GBFS content, the Ca/Si ratio in C-A-S-H gel gradually reduces, while the Al/Si ratio increases. In general, Al tetrahedrons are substituted into the paired sites of Si-O tetrahedron chains, and Si or Al tetrahedrons are connected to form a frame structure by sharing one oxygen atom in C-(A)-S-H gels [50]. The composition of C-A-S-H in alkali-activated material relies on the calcium and aluminum contents of precursors [50]. Therefore, the change in ratios is related to the higher aluminum content in GBFS. In relation to pore structure, compared to Portland cement, alkali-activated SGM has a lower total porosity and a higher number of gel pores but fewer capillary pores. This may be due to the higher content of C-A-S-H gel in alkali-activated SGM compared to C-S-H gel in Portland cement [54–60]. Compared to alkali-activated GBFS material, alkali-activated SGM has a higher porosity and fewer gel pores. Similarly, compared to alkali-activated steel slag material, alkali-activated SGM exhibits a reduced content of small pores and a decreasing trend in porosity with the increase in GBFS content, indicating that GBFS plays a role in refining the pore structure [54–60].

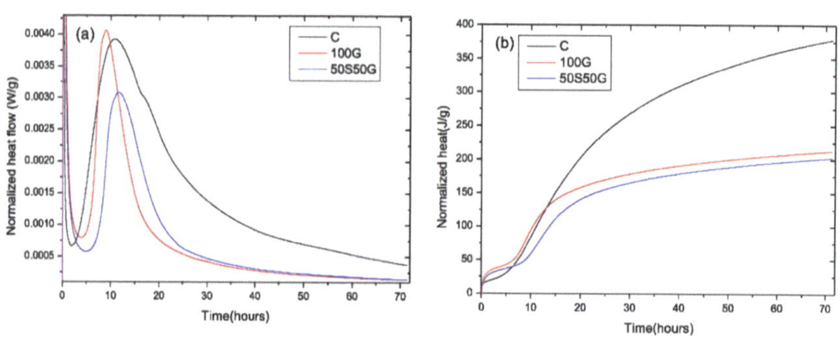

Figure 6. Hydration heat evolution curves of alkali-activated cementitious material and Portland cement: (**a**) exothermic rate; (**b**) cumulative hydration heat. (C: cement; 100G: 100% GBFS; 50S50G: GBFS: steel slag = 1:1) [47].

3. Fresh Properties of SGM

3.1. *Fluidity*

There is limited research specifically focusing on the fluidity of binary systems of SGM, as well as ternary systems of SGM containing gypsum. However, there are studies regarding the fluidity of SGM modified with alkaline activators. It has been found that when sodium silicate is used as the activator, the flowability of SGM with a mass ratio (steel slag/GBFS) of 1:1 and 0:1 is reported to be 230 mm and 270 mm, respectively. The specific surface areas of steel slag and GBFS used in these systems are recorded as 370 m^2/kg and 436 m^2/kg [49]. For alkali-activated SGM with mass ratios of 8:2, 9:1, and 10:0 and specific surface areas of 458 m^2/kg and 430 m^2/kg for steel slag and GBFS, the flowability is measured as 257 mm, 249 mm, and 234 mm, respectively [59]. Generally, under the same alkaline environment and normal curing conditions, the fluidity of alkali-activated SGM is better than that of alkali-activated GBFS binder and alkali-activated steel slag binder. This is due to the higher reactivity of GBFS compared to SGM in the same alkaline environment. It is also noted to be slightly better than that of Portland cement.

3.2. *Setting Time*

The reaction of SGM is typically slow, leading to a relatively long setting time of the paste. Tsai et al. [51] observed that the relative content of steel slag and GBFS has a significant impact on the setting time of SGM in systems. Under standard curing conditions, when the mass ratio of steel slag to GBFS is 1:9, the setting time can reach 4 d. On the other hand, when the ratio is below 7:3, the setting time is reduced to only 0.5 h. This indicates that as the steel slag content increases, the setting time of SGM reduces. This observation can be attributed to the fact that higher steel slag content more effectively promotes the pozzolanic reaction of GBFS.

In the system of SGM with cement clinker, the research finished by Zhao et al. [57] shows that under standard curing conditions, when the relative content of steel slag, GBFS, and cement clinker is 35:35:30 and the water-to-binder ratio is 0.35 the initial and final setting times of SGM (sample RBC) are 250 min and 325 min, respectively. These setting times significantly exceed those of Portland cement (sample PC). Additionally, when the raw materials meet the Fuller particle distribution (samples F-1#–F-5#), the initial and final setting times are significantly reduced but still remain longer than the setting time of Portland cement (Figure 7).

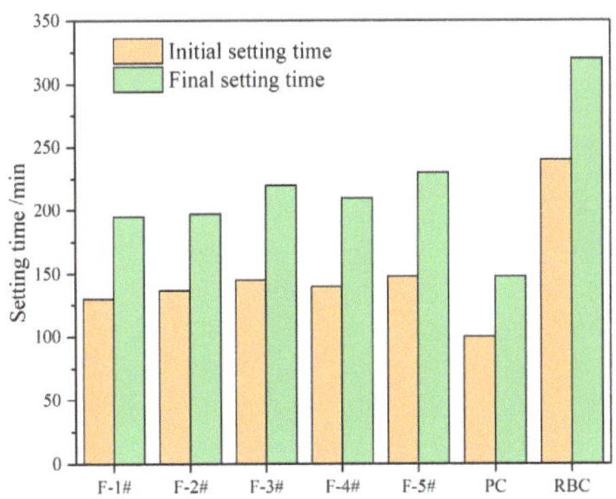

Figure 7. Setting times of pastes [57].

When alkaline activators are used as modifiers in SGM, the setting time is further reduced compared to cement-modified binder systems. This is due to the higher alkaline environment provided by the alkaline activators. You et al. [47] found that the setting time of water glass-activated SGM is only 50 min, which is much shorter than that of cement. Moreover, with the increase in steel slag content in SGM, there is a slight prolongation of the initial setting time. Some scholars believe that this can be due to the lower solubility of steel slag compared to GBFS [61–64]. The presence of a large number of inert components, primarily in the form of the RO phase in steel slag, reduces the content of active components. These inert components also do not dissolve in an alkaline environment [61–68].

It is important to note that while alkaline activators can significantly reduce the setting time of SGM compared to cement-modified systems, the setting time may vary depending on the composition and properties of the SGM mixture. Further research is needed to investigate and optimize the setting time of SGM in different formulations to meet specific application requirements.

4. Hardened Properties of SGM

4.1. Mechanical Strength

In the binary composite system of SGM, the early reaction rate of steel slag and GBFS is extremely slow, leading to a severe lack of early strength. However, when modified with gypsum, SGM exhibits improved early and later strength. With a mass ratio of 2:7 for steel slag and GBFS, gypsum content of 12%, and a sand ratio of 0.42, the 1 d compressive strength of concrete cured at 45 °C can reach 24.06 MPa, and the 28 d compressive strength can reach 51.54 MPa [68]. The strength of SGM is influenced by the relative content of each component. Cui et al. [17] conducted a study on the influence of steel slag content on the compressive strength of SGM at 3 d, 7 d, and 28 d under standard curing conditions, as depicted in Figure 8. The results show that increasing the relative content of steel slag has little effect on the compressive strength at 7 d and 28 d, but it gradually reduces the compressive strength at 3 d, and the decrease becomes more significant when the steel slag content exceeds 30%. This is because in the early hydration stage, the insufficient filling effect of steel slag as a micro-aggregate cannot compensate for its low reactivity, and with a higher content of steel slag, the low content of GBFS cannot provide enough amorphous active components such as aluminosilicate, which affects the hydration synergy among steel slag, GBFS, and gypsum. Other scholars focus on the influence of GBFS content on the strength of SGM. Zhang et al. [63] found that under standard curing conditions, the compressive strength at different periods is the highest when the relative content of steel slag, GBFS, and gypsum is 29:58:13. However, as the GBFS content increases, the decrease in later strength becomes more pronounced. In addition, the dosage of gypsum, as an important modifier, also has limitations. Xu et al. [69] discovered that raising the curing temperature appropriately helps improve the mechanical properties of the binder at all ages. The higher the temperature, the higher the optimal relative content of steel slag. Under the same mix proportion, the 1 d and 28 d compressive strength of concrete with a mass ratio of steel slag to GBFS of 0.55, gypsum content of 15%, and a sand ratio of 0.43 cured at 45 °C are 27.23 MPa and 49 MPa, respectively. These are significantly higher than the 1 d and 28 d compressive strength of concrete cured at 20 °C, with an increase of approximately 22.15 MPa and 7 MPa, respectively [69]. This is because the higher temperature facilitates the dissolution and deflocculation of active components in steel slag and GBFS [70–74].

When a small amount of cement clinker is used as a modifier, the later compressive strength of SGM is comparable to that of the Portland cement system, but its early strength benefits from the disaggregation of the glassy phase in GBFS and the dissolution of active components in steel slag, thereby improving the strength at all ages. However, the higher content of clinker has a negative effect on the hydration of SGM [49]. Li et al. [48] found that $Ca(OH)_2$ formed by the hydration of cement increases the alkalinity of the liquid phase. However, if the content of cement is too high, it can actually decrease the strength of the system. This is because when there is an excess amount of cement, the concentration of Ca^{2+} in the early solution quickly reaches saturation, making it difficult for it to react with GBFS in a timely manner, impeding the dissolution of calcium aluminate and calcium silicate in steel slag [75–77].

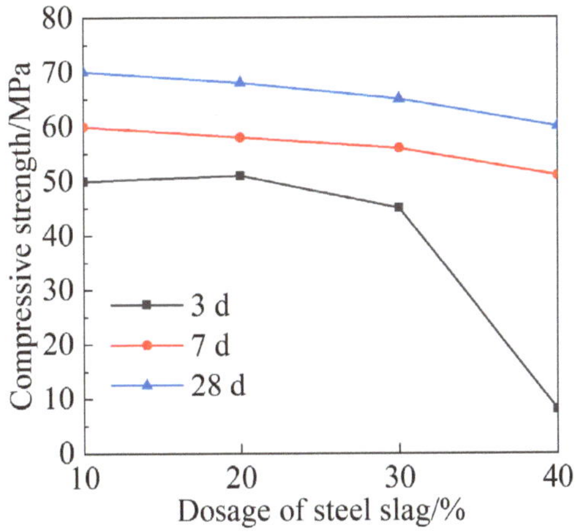

Figure 8. Influence of steel slag content on compressive strength of SGM with gypsum system [17].

Providing an alkaline environment directly is the easiest way to improve the mechanical properties. Research has shown that several commonly used alkaline activators have different stimulating effects on SGM, with water glass having the best effect, followed by NaOH, Na_2SO_4, and Na_2CO_3 having the worst effect [34,77]. When water glass is used as the alkaline activator, the 28 d compressive strength can reach 76.6 MPa thanks to the dual stimulation of OH^- and $[SiO_4]^{4-}$. However, when the water glass content is too high, the strength actually decreases. This is because the rapidly generated hydration products cannot diffuse in time and adhere to the surface of steel slag and GBFS particles, hindering further hydration [57]. In contrast to the results of cement modification, when water glass is used as an activator, the compressive strength of SGM at different ages can be comparable to or even slightly better than that of Portland cement [57]. In an alkaline environment, increasing the relative content of steel slag will decrease the compressive strength of the system [57,63,78–80]. In experiments conducted by You et al. [47], when water glass is used as an alkaline activator, the 1 d and 28 d compressive strength of alkali-activated GBFS material (sample 100G) under standard conditions are 37 MPa and 72.5 MPa, respectively. After replacing 50% of GBFS with steel slag (sample 50S50G), the compressive strength at 1 d and 28 d decreases by about 28% and 10%, respectively, but still reaches the strength level of the Portland cement system (Figure 9). In addition to the alkaline activators containing Na, some activators containing Ca, such as CaO and $Ca(OH)_2$, can also have good activation effects. These activators not only provide an alkaline environment but also provide more Ca for hydration. In this case, the effect of the activators is more similar to that of clinker [80,81]. Wang et al. [78] added a small amount of $Ca(OH)_2$ to the water glass-activated SGM and found that with only a 2% addition of $Ca(OH)_2$, the compressive strength at 1 d and 28 d reached approximately 57 MPa and 89 MPa, respectively. Compared to the system without $Ca(OH)_2$, the compressive strength increases by about 27% and 25%, respectively. This is mainly attributed to the large amount of active Ca^{2+} provided by $Ca(OH)_2$, which promotes the formation of silicate and aluminosilicate networks and provides nucleation sites for C-S-H gel, thus promoting hydration.

In summary, while the early compressive strength is a key performance indicator of SGM and has received much attention, research on mechanical properties such as tensile strength, flexural strength, and elastic modulus is still lacking, and there is still insufficient emphasis on the later strength.

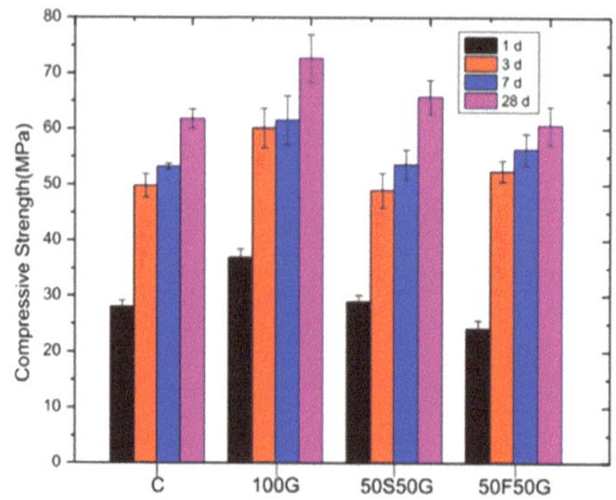

Figure 9. Mechanical properties of alkali-activated material and cement [58].

4.2. Durability

Currently, there is no research specifically focused on the durability of the binary system of SGM, the ternary system of SGM with gypsum, or the ternary system of SGM with cement clinker. However, there have been some studies on the durability of SGM under alkaline conditions.

Regarding water resistance, You et al. [47] compared the water absorption of Portland cement mortars, alkali-activated SGM mortars, and alkali-activated GBFS mortars with the same strength grade, as shown in Figure 10. They found that alkali-activated materials have lower water absorption compared to Portland cement of the same strength grade, and the water absorption of alkali-activated SGM is slightly higher than that of alkali-activated GBFS material, indicating that increasing the steel slag content increases the water absorption, which in turn increases the risk of concrete moisture content, deformation, and cracking.

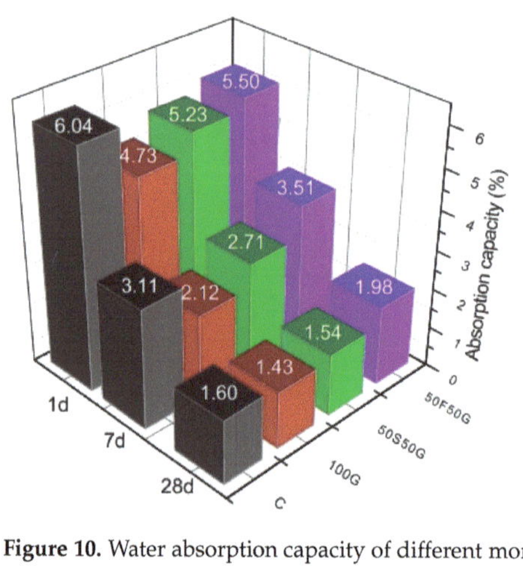

Figure 10. Water absorption capacity of different mortars at the same strength level [47].

In terms of frost resistance, Xiang et al. [76] observed that the average dynamic modulus of elasticity of alkali-activated SGM decreased to 66% of the original value after 300 cycles of freeze–thaw. The mass loss is only 1.1% and the frost resistance reaches the F300 level. Li et al. [77] compared the durability of alkali-activated steel slag material with Portland cement and found that the frost resistance of alkali-activated steel slag cementitious materials was better than that of Portland cement. The compressive strength loss rate and mass loss rate of alkali-activated steel slag cementitious materials after 50 freeze–thaw cycles are 5.9% and 1.03%, respectively, lower than the values of 9.6% and

1.11% for Portland cement. This is due to the low pore volume and high content of gel pore in the hardened paste of alkali-activated SGM, resulting in a denser pore structure with fewer interconnected cracks formed during freeze–thaw cycles [38].

In terms of carbonation resistance, Xiang et al. [76] observed that the carbonation depth of alkali-activated SGM was only 0.3 mm after 28 d under the condition of 20 ± 3% CO_2 concentration and 70 ± 5% relative humidity, indicating a negligible level of carbonation.

In terms of resistance to chloride ion penetration, Xiang et al. [76] conducted tests using the electrical flux method and found that the average charge passed within 6 h for alkali-activated SGM was less than 1000 C, indicating very low permeability to chloride ions. You et al. [47] compared the chloride ion resistance of alkali-activated SGM with Portland cement using the rapid chloride migration coefficient method and found that alkali-activated SGM exhibited better resistance to chloride ion penetration, with a chloride ion diffusion coefficient of only one-seventh of that of Portland cement. This is mainly due to the lower interconnected pore volume in alkali-activated SGM compared to Portland cement and the formation of C-A-S-H gel as the main hydration product in alkali-activated SGM, which results in slower chloride ion migration compared to C-S-H gel [78–80].

Overall, compared to Portland cement, SGMs with alkaline activators demonstrate excellent water resistance, frost resistance, carbonation resistance, and resistance to chloride ion penetration. This provides a solid foundation for their design and application in long-life and highly durable materials.

4.3. Shrinkage

Shrinkage may lead to internal tensile stresses and bring a risk of cracking, causing structural instability [80–82]. Currently, there is limited research on the volume stability of SGM, SGM with gypsum, and SGM with clinker. However, there is a scarcity of studies specifically investigating the volume stability of alkali-activated SGM. Research conducted by You et al. [47] indicated that the autogenous and drying shrinkage of alkali-activated SGM falls between those of Portland cement and alkali-activated GBFS cementitious materials (Figure 11). It should be noted that alkali-activated GBFS cementitious materials tend to exhibit high shrinkage, which can be partially mitigated by adding steel slag. However, even with the addition of steel slag, the shrinkage of alkali-activated SGM remains higher than that of cement. In terms of autogenous shrinkage, the autogenous shrinkage of alkali-activated SGM mainly occurs within the first 3 d and is approximately 0.7 times that of alkali-activated GBFS cementitious materials and twice that of Portland cement. Concerning drying shrinkage of alkali-activated SGM, it develops rapidly in the early stages, specifically before 28 d, and then slows down. The drying shrinkage at 28 d is approximately 0.7 times that of alkali-activated GBFS cementitious materials but still much higher than that of Portland cement. This is attributed to the main hydration product of alkali-activated GBFS cementitious materials, C-A-S-H gel, which experiences significant shrinkage under dry conditions [83–85]. The inclusion of low-reactivity steel slag in the system acts as a micro-aggregate, partially inhibiting the drying shrinkage of alkali-activated SGM [86]. Research by Sun et al. [50] also demonstrated that the drying shrinkage of alkali-activated steel slag cementitious materials decreases with increasing content of steel slag. For instance, when steel slag content reaches 30%, the 180 d drying shrinkage value is 3070×10^{-6}, but when the steel slag content is 70%, the 180 d drying shrinkage value drops to only 490×10^{-6}, reducing by approximately six times. This suggests that increasing the steel slag content significantly improves the drying shrinkage of alkali-activated steel slag cementitious materials.

Overall, the shrinkage of alkali-activated steel slag cementitious materials is still higher than that of Portland cement, and further research is needed to investigate the shrinkage effects of SGM with gypsum and clinker, as well as the measures to mitigate shrinkage in alkali-activated systems.

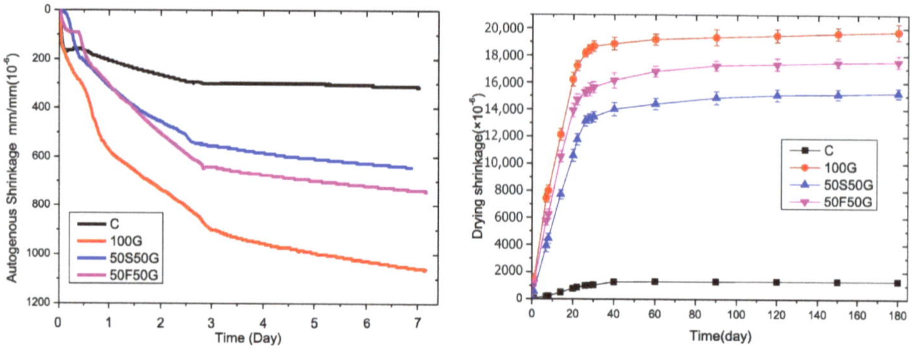

Figure 11. Autogenous shrinkage and drying shrinkage of cementitious materials [47].

5. Conclusions and Outlook

While steel slag and GBFS can exhibit some degree of synergistic hydration in water, the slow pozzolanic reaction of GBFS is attributed to the weak alkalinity generated by steel slag. To enhance the hydration rate, shorten the setting time, and improve the early strength, the addition of gypsum-based materials (such as natural gypsum and by-product gypsum), cement clinker, or alkaline activators (such as sodium silicate, $Ca(OH)_2$, NaOH, Na_2CO_3, etc.) can effectively optimize the SGM. Alkaline activators and clinker promote the hydration of steel slag and GBFS by increasing the alkalinity of the reaction system. In the case of SGM with gypsum, there is a clear ternary synergistic effect, leading to the formation of a dense matrix with AFt and C-S-H gel as the main reaction products. Despite the attention given to the hydration and hardening mechanisms, as well as the performance improvement mechanisms of SGM, there are still some unresolved issues:

(1) A common issue with SGM is that as the steel slag content increases, the setting time lengthens and mechanical properties at an early age decrease. Given the high utilization rate and cost of GBFS in the market, as well as the low utilization rate and cost of steel slag, further research is needed to maximize the utilization of steel slag, reduce raw material costs, and ensure critical early-age performance.

(2) There is limited research on the fresh properties of SGM. In practical engineering, fresh pastes need to meet certain requirements for setting time, workability, pumpability, and water retention. Therefore, identifying the factors and key indicators affecting fresh properties of SGM, as well as developing effective measures to improve them, requires further investigation.

(3) SGM with gypsum holds promise as a solid waste system. However, there is still a lack of systematic research on the long-term mechanical properties, durability, and volume stability. In actual service environments, the secondary hydration of gypsum and the occurrence of "delayed AFt" may risk cracking in hardened pastes. Additionally, the presence of gypsum may lead to poor water resistance of the cementitious materials, necessitating further investigation.

Author Contributions: Conceptualization, methodology, writing—original draft, funding acquisition, J.S.; investigation, validation, formal analysis, resources, S.H.; methodology, investigation, Y.G.; investigation, formal analysis, X.C.; data curation, D.Z. All authors have read and agreed to the published version of the manuscript.

Funding: This research was funded by the Natural Science Fund of Shandong Province, ZR2023QE007, and the Open Project of Engineering Research Center of Concrete Technology under Marine Environment, TMduracon2022009.

Data Availability Statement: The raw data supporting the conclusions of this article will be made available by the authors on request.

Conflicts of Interest: Author Dongdong Zhang was employed by the company Jinan Rail Transit Group Co., Ltd. The remaining authors declare that the research was conducted in the absence of any commercial or financial relationships that could be construed as a potential conflict of interest.

References

1. Li, J.; Ni, W.; Wang, X.; Zhu, S.; Wei, X.; Jiang, F.; Zeng, H.; Hitch, M. Mechanical activation of medium basicity steel slag under dry condition for carbonation curing. *J. Build. Eng.* **2022**, *50*, 104123. [CrossRef]
2. Gao, W.; Zhou, W.; Lyu, X.; Liu, X.; Su, H.; Li, C.; Wang, H. Comprehensive utilization of steel slag: A review. *Powder Technol.* **2023**, *422*, 118449. [CrossRef]
3. Guo, J.L.; Bao, Y.P.; Wang, M. Steel slag in China, Treatment, recycling, and management. *Waste Manag.* **2018**, *78*, 318–330. [CrossRef] [PubMed]
4. Yang, M.Q.; Yang, J.Y. Vanadium extraction from steel slag: Generation, recycling and management. *Environ. Pollut.* **2024**, *343*, 123126. [CrossRef]
5. Aliyah, F.; Kambali, I.; Setiawan, A.F.; Radzi, Y.M.; Rahman, A.A. Utilization of steel slag from industrial waste for ionizing radiation shielding concrete: A systematic review. *Constr. Build. Mater.* **2023**, *382*, 131360. [CrossRef]
6. Kurniati, E.O.; Pederson, F.; Kim, H.-J. Application of steel slags, ferronickel slags, and copper mining waste as construction materials: A review. *Resour. Conserv. Recycl.* **2023**, *198*, 107175. [CrossRef]
7. Shu, K.; Sasaki, K. Occurrence of steel converter slag and its high value-added conversion for environmental restoration in China: A review. *J. Clean. Prod.* **2022**, *373*, 133876. [CrossRef]
8. O'Connor, J.; Nguyen, T.B.T.; Honeyands, T.; Monaghan, B.; O'Dea, D.; Rinklebe, J.; Vinu, A.; Hoang, S.A.; Singh, G.; Kirkham, M.B.; et al. Production, characterisation, utilisation, and beneficial soil application of steel slag: A review. *J. Hazard Mater.* **2021**, *419*, 126478. [CrossRef] [PubMed]
9. Tang, S.W.; Cai, R.J.; He, Z.; Cai, X.H.; Shao, H.Y.; Li, Z.J.; Yang, H.M.; Chen, E. Continuous microstructural correlation of slag/superplasticizer cement pastes by heat and impedance methods via fractal analysis. *Fractals* **2017**, *25*, 1740003. [CrossRef]
10. Zhuang, S.Y.; Wang, Q. Inhibition mechanisms of steel slag on the early-age hydration of cement. *Cem. Concr. Res.* **2021**, *140*, 106283. [CrossRef]
11. Yang, H.M.; Zhang, S.M.; Wang, L.; Chen, P.; Shao, D.K.; Tang, S.W.; Li, J.Z. High-ferrite Portland cement with slag: Hydration, microstructure, and resistance to sulfate attack at elevated temperature. *Cem. Concr. Comp.* **2022**, *130*, 104560–104576. [CrossRef]
12. Luo, T.; Wang, X.; Zhuang, S. Value-added utilization of steel slag as a hydration heat controlling material to prepare sustainable and green mass concrete. *Case Stud. Constr. Mater.* **2023**, *19*, e02619. [CrossRef]
13. Zhuang, S.Y.; Wang, Q.; Luo, T. Effect of C12A7 in steel slag on the early-age hydration of cement. *Cem. Concr. Res.* **2022**, *162*, 107010. [CrossRef]
14. Wang, Q.; Yan, P.Y. Hydration properties of basic oxygen furnace steel slag. *Constr. Build. Mater.* **2010**, *24*, 1134–1140. [CrossRef]
15. Sun, J.W.; Hou, S.Y.; Guo, Y.H. Effects of high-temperature curing on hydration and microstructure of alkali-activated typical steel slag cementitious material. *Dev. Built Environ.* **2024**, *17*, 100314. [CrossRef]
16. Li, Y.; Wu, B.H.; Ni, W.; Mu, X.L. Synergies in early hydration reaction of slag-steel slag-gypsum system. *J. Northeast. Univ.* **2020**, *41*, 581–586.
17. Cui, X.W.; Ni, W.; Reng, C. Chemical activation of cementitious materials with all solid waste based of steel slag and blast furnace slag. *Chin. J. Mater. Res.* **2017**, *9*, 687–694.
18. Wang, Q.; Yan, P.Y.; Feng, J.W. A discussion on improving hydration activity of steel slag by altering its mineral compositions. *J. Hazard. Mater.* **2011**, *186*, 1070–1075. [CrossRef] [PubMed]
19. Guo, X.L.; Shi, H.S. Effects of steel slag admixture with GGBFS on performances of cement paste and mortar. *Adv. Cem. Res.* **2015**, *26*, 93–100. [CrossRef]
20. Wang, Q.; Yan, P.Y.; Mi, G.D. Effect of blended steel slag-GBFS mineral admixture on hydration and strength of cement. *Constr. Build. Mater.* **2012**, *35*, 8–14. [CrossRef]
21. Singh, S.; Tripathy, D.; Ranjith, P. Performance evaluation of cement stabilized fly ash–GBFS mixes as a highway construction material. *Waste Manag.* **2008**, *28*, 1331–1337. [CrossRef]
22. Chen, P.; Ma, B.; Tan, H.; Wu, L.; Zheng, Z.; He, X.; Li, H.; Jin, Z.; Li, M.; Lv, Z. Improving the mechanical property and water resistance of β-hemihydrate phosphogypsum by incorporating ground blast-furnace slag and steel slag. *Constr. Build. Mater.* **2022**, *344*, 128265. [CrossRef]
23. Zeng, Q.; Liu, X.; Zhang, Z.; Xu, C.C. Synergistic utilization of blast furnace slag with other industrial solid wastes in cement and concrete industry, Synergistic mechanisms, applications, and challenges. *Green Energy Resour.* **2023**, *1*, 100012. [CrossRef]
24. Zhang, M.; Li, K.; Ni, W.; Zhang, S.; Liu, Z.; Wang, K.; Wei, X.; Yu, Y. Preparation of mine backfilling from steel slag-based non-clinker combined with ultra-fine tailing. *Constr. Build. Mater.* **2022**, *320*, 126248. [CrossRef]
25. Wang, Q.; Miao, M.; Feng, J.; Yan, P. The influence of high-temperature curing on the hydration characteristics of a cement–GGBS binder. *Adv. Cem. Res.* **2012**, *24*, 33–40. [CrossRef]

26. Xiao, B.; Wen, Z.; Miao, S.; Gao, Q. Utilization of steel slag for cemented tailings backfill: Hydration, strength, pore structure, and cost analysis. *Case Stud. Constr. Mater.* **2021**, *15*, e00621. [CrossRef]
27. Huang, X.; Wang, Z.; Liu, Y.; Hu, W.; Ni, W. On the use of blast furnace slag and steel slag in the preparation of green artificial reef concrete. *Constr. Build. Mater.* **2016**, *112*, 241–246. [CrossRef]
28. Wang, Q.; Yang, J.W.; Yan, P.Y. Cementitious properties of super-fine steel slag. *Powder Technol.* **2013**, *245*, 35–39. [CrossRef]
29. Zhou, Y.; Li, W.; Peng, Y.; Tang, S.; Wang, L.; Shi, Y.; Li, Y.; Wang, Y.; Geng, Z.; Wu, K. Hydration and fractal analysis on Low-heat Portland cement pastes by thermodynamic-based methods. *Fractal Fract.* **2023**, *7*, 606. [CrossRef]
30. Hamdan, A.; Song, H.; Yao, Z.; Alnahhal, M.F.; Kim, T.; Hajimohammadi, A. Modifications to reaction mechanisms, phase assemblages and mechanical properties of alkali-activated slags induced by gypsum addition. *Cem. Concr. Res.* **2023**, *174*, 107311. [CrossRef]
31. Li, Z.; Lu, T.; Liang, X.; Dong, H.; Ye, G. Mechanisms of autogenous shrinkage of alkali-activated slag and fly ash pastes. *Cem. Concr. Res.* **2020**, *135*, 106107. [CrossRef]
32. Fu, Q.; Bu, M.; Zhang, Z.; Xu, W.; Yuan, Q.; Niu, D. Hydration characteristics and microstructure of alkali-activated slag concrete: A review. *Engineering* **2023**, *20*, 162–179. [CrossRef]
33. Lee, K.M.; Park, P.J. Estimation of the environmental credit for the recycling of granulated blast furnace slag based on LCA. *Resour. Conserv. Recycl.* **2005**, *44*, 139–151. [CrossRef]
34. Zhao, J.; Li, Z.; Wang, D.; Yan, P.; Luo, L.; Zhang, H.; Zhang, H.; Gu, X. Hydration superposition effect and mechanism of steel slag powder and granulated blast furnace slag powder. *Constr. Build. Mater.* **2023**, *336*, 130101. [CrossRef]
35. Geng, Z.; Tang, S.; Wang, Y.; He, Z.; Wu, K. The stress relaxation properties of calcium silicate hydrate: A molecular dynamics study. *J. Zhejiang Univ.-Sci. A* **2023**, 1–22. [CrossRef]
36. Liao, Y.; Wang, S.; Wang, K.; Al Qunaynah, S.; Wan, S.; Yuan, Z.; Xu, P.; Tang, S. A study on the hydration of calcium aluminate cement pastes containing silica fume using non-contact electrical resistivity measurement. *J. Mater. Res. Technol.* **2023**, *24*, 8135–8149. [CrossRef]
37. Wang, L.; Guo, F.; Lin, Y.; Yang, H.; Tang, S.W. Comparison between the effects of phosphorous slag and fly ash on the C-S-H structure, long-term hydration heat and volume deformation of cement-based materials. *Constr. Build. Mater.* **2020**, *250*, 118807. [CrossRef]
38. Wang, L.; Jin, M.; Zhou, S.; Tang, S.; Lu, X. Investigation of microstructure of C-S-H and micro-mechanics of cement pastes under NH4NO3 dissolution by 29Si MAS NMR and microhardness. *Measurement* **2021**, *185*, 110019. [CrossRef]
39. Wang, L.; Zhou, S.; Shi, Y.; Huang, Y.; Zhao, F.; Huo, T.; Tang, S. The influence of fly ash dosages on the permeability, pore structure and fractal features of face slab concrete. *Fractal Fract.* **2022**, *6*, 476. [CrossRef]
40. Wang, L.; Huang, Y.; Zhao, F.; Huo, T.; Chen, E.; Tang, S. Comparison between the influence of finely ground phosphorous slag and fly ash on frost resistance, pore structures and fractal features of hydraulic concrete. *Fractal Fract.* **2022**, *6*, 598. [CrossRef]
41. Tsai, C.J.; Huang, R.; Lin, W.T.; Chiang, H.W. Using GGBOS as the alkali activators in GGBS and GGBOS blended cements. *Constr. Build. Mater.* **2014**, *70*, 501–507. [CrossRef]
42. Meshram, S.; Raut, S.P.; Ansari, K.; Madurwar, M.; Daniyal, M.; Khan, M.A.; Katare, V.; Khan, A.H.; Khan, N.A.; Hasan, M.A. Waste slags as sustainable construction materials, a compressive review on physico mechanical properties. *J. Mate. Res. Technol.* **2023**, *23*, 5821–5845. [CrossRef]
43. Zhao, J. Grinding and Hydration Characteristics of Steel Slag and Composition and Properties of Composite Cementitious Materials Containing Steel. Ph.D. Thesis, China University of Mining and Technology, Beijing, China, 2015.
44. Xiong, X.; Yang, Z.; Yan, X.; Zhang, Y.; Dong, S.; Li, K.; Briseghella, B.; Marano, G.C. Mechanical properties and microstructure of engineered cementitious composites with high volume steel slag and GGBFS. *Constr. Build. Mater.* **2014**, *70*, 501–507. [CrossRef]
45. Ghorbani, S.; Stefanini, L.; Sun, Y.; Walkley, B.; Provis, J.L.; De Schutter, G.; Matthys, S. Characterisation of alkali-activated stainless steel slag and blast-furnace slag cements. *Cem. Concr. Compos.* **2023**, *143*, 105230. [CrossRef]
46. Siddique, R.; Bennacer, R. Use of iron and steel industry by-product (GGBS) in cement paste and mortar. *Resour. Conserv. Recycl.* **2012**, *69*, 29–34. [CrossRef]
47. You, N.; Li, B.; Cao, R.; Shi, J.; Chen, C.; Zhang, Y. The influence of steel slag and ferronickel slag on the properties of alkali-activated slag mortar. *Constr. Build. Mater.* **2019**, *227*, 116614. [CrossRef]
48. Li, J.S.; Li, J.M.; Ge, X.X. Preparation and characterization of early strengthened of binding materials steel slag and blast furnace slag. *J. Anhui Univ. Technol. (Nat. Sci.)* **2020**, *37*, 321–326.
49. Zhou, Y.Q.; Sun, J.; Liao, Y.W. Influence of ground granulated blast furnace slag on the early hydration and microstructure of alkali-activated converter steel slag binder. *J. Therm. Anal. Calorim.* **2020**, *182*, 243–252. [CrossRef]
50. Sun, J.W.; Chen, Z.H. Effect of silicate modulus of water glass on the hydration of alkali-activated converter steel slag. *J. Therm. Anal. Calorim.* **2019**, *138*, 47–56. [CrossRef]
51. Atiş, C.D.; Bilim, C.; Çelik, Ö.; Karahan, O. Influence of activator on the strength and drying shrinkage of alkali-activated slag mortar. *Constr. Build. Mater.* **2009**, *23*, 548–555. [CrossRef]
52. Luukkonen, T.; Sreenivasan, H.; Abdollahnejad, Z.; Yliniemi, J.; Kantola, A.; Telkki, V.V.; Kinnunen, P.; Illikainen, M. Influence of sodium silicate powder silica modulus for mechanical and chemical properties of dry-mix alkali-activated slag mortar. *Constr. Build. Mater.* **2020**, *233*, 117354. [CrossRef]

53. Liu, Z.; Ni, W.; Li, Y.; Ba, H.; Li, N.; Ju, Y.; Zhao, B.; Jia, G.; Hu, W. The mechanism of hydration reaction of granulated blast furnace slag-steel slag-refining slag-desulfurization gypsum-based clinker-free cementitious materials. *J. Build. Eng.* **2021**, *40*, 103289. [CrossRef]
54. Li, Y.; Liang, W.; Ni, W.; Mu, X.; Li, Y.; Fan, K. Characteristics of hydration and hardening of steel slag mud-blast furnace slag-desulphurization gypsum system. *Bull. Chin. Ceram. Soc.* **2022**, *41*, 536–544.
55. Liao, Y.; Yao, J.; Deng, F.; Li, H.; Wang, K.; Tang, S. Hydration behavior and strength development of supersulfated cement prepared by calcined phosphogypsum and slaked lime. *J. Build. Eng.* **2023**, *80*, 108075. [CrossRef]
56. Xu, C.; Ni, W.; Li, K.; Zhang, S.; Xu, D. Activation mechanisms of three types of industrial by-product gypsums on steel slag-granulated blast furnace slag-based binders. *Constr. Build. Mater.* **2021**, *288*, 123111. [CrossRef]
57. Zhao, J.H.; Wang, D.M.; Yan, P.Y. Design and experimental study of a ternary blended cement containing high volume steel slag and blast-furnace slag based on Fuller distribution model. *Constr. Build. Mater.* **2017**, *140*, 248–256. [CrossRef]
58. Li, Y.; Zhang, H.; Huang, M.; Yin, H.; Jiang, K.; Xiao, K.; Tang, S. Influence of different alkali sulfates on the shrinkage, hydration, pore structure, fractal dimension and microstructure of low-heat Portland cement, medium-heat Portland cement and ordinary Portland cement. *Fractal Fract.* **2021**, *5*, 79. [CrossRef]
59. Feng, J.J.; Sun, J.W. A comparison of the 10-year properties of converter steel slag activated by high temperature and an alkaline activator. *Constr. Build. Mater.* **2020**, *234*, 116948. [CrossRef]
60. Maria, C.; Willian, A.; Isabel, S. Microstructural and mechanical properties of alkali activated colombian raw materials. *Materials* **2016**, *9*, 158.
61. Sun, J.; Zhang, Z.; Zhuang, S.; He, W. Hydration properties and microstructure characteristics of alkali-activated steel slag. *Constr. Build. Mater.* **2020**, *241*, 118141. [CrossRef]
62. Cao, R.; Li, B.; You, N.; Zhang, Y.; Zhang, Z. Properties of alkali-activated ground granulated blast furnace slag blended with ferronickel slag. *Constr. Build. Mater.* **2018**, *192*, 123–132. [CrossRef]
63. Zhang, G.Q.; Wu, P.C.; Gao, S.J. Properties and microstructure of low-carbon whole-tailings cemented paste backfill material containing steel slag, granulated blast furnace slag and flue gas desulphurization gypsum. *Acta Microsc.* **2019**, *28*, 770–780.
64. Brakat, A.; Zhang, Y. Shrinkage mitigation of alkali activated slag with natural cellulose fibers. *Adv. Cem. Res.* **2017**, *31*, 47–57. [CrossRef]
65. Furlani, E.; Maschio, S.; Magnan, M.; Aneggi, E.; Andreatta, F.; Lekka, M.; Lanzutti, A.; Fedrizzi, L. Synthesis and characterization of geopolymers containing blends of unprocessed steel slag and metakaolin: The role of slag particle size. *Ceram. Int.* **2018**, *44*, 5226–5232. [CrossRef]
66. Hu, S.G.; He, Y.J.; Lu, L.N.; Ding, Q.J. Effect of fine steel slag powder on the early hydration process of Portland cement. *J. Wuhan Univ. Technol. (Mater. Sci. Ed.)* **2006**, *21*, 147–149.
67. Altun, I.A.; Yılmaz, İ. Study on steel furnace slags with high MgO as additive in Portland cement. *Cem. Concr. Res.* **2002**, *32*, 1247–1249. [CrossRef]
68. Liu, K.S.; Zhang, Z.Q.; Sun, J.W. Advances in understanding the alkali-activated metallurgical slag. *Adv. Civ. Eng.* **2021**, *2021*, 8795588. [CrossRef]
69. Xu, C.; Ni, W.; Li, K.; Zhang, S.; Li, Y.; Xu, D. Hydration mechanism and orthogonal optimisation of mix proportion for steel slag-slag-based clinker-free prefabricated concrete. *Constr. Build. Mater.* **2019**, *228*, 117036. [CrossRef]
70. Xiao, B. Steel Slag Binder and Its Application in the Cemented Tailings Backfill. Ph.D. Thesis, University of Science and Technology, Beijing, China, 2020.
71. Adesanya, E.; Ohenoja, K.; Di Maria, A.; Kinnunen, P.; Illikainen, M. Alternative alkali-activator from steel-making waste for one-part alkali-activated slag. *J. Clean. Prod.* **2020**, *274*, 123020. [CrossRef]
72. Han, F.H.; Yan, P.Y. Hydration characteristics of slag-blended cement at different temperatures. *J. Sustain. Cem.-Based Mater.* **2015**, *4*, 34–43. [CrossRef]
73. Clark, B.A.; Brown, P.W. Formation of calcium sulfoaluminate hydrate compounds. *Cem. Concr. Res.* **2000**, *30*, 233–240. [CrossRef]
74. Grusczscinski, E.; Brown, P.W.; Both, J.V., Jr. The formation of ettringite at elevated temperature. *Cem. Concr. Res.* **1993**, *23*, 981–987. [CrossRef]
75. Wang, M.; Qian, B.; Jiang, J.; Liu, H.; Cai, Q.; Ma, B.; Hu, Y.; Wang, L. The reaction between Ca^{2+} from steel slag and granulated blast-furnace slag system, a unique perspective. *Chem. Pap.* **2020**, *74*, 4401–4410. [CrossRef]
76. Xiang, X.D.; Xi, J.C.; Li, C.H.; Jiang, X.W. Preparation and application of the cement-free steel slag cementitious material. *Constr. Build. Mater.* **2016**, *114*, 874–879. [CrossRef]
77. Li, Q.; Zhao, F.Q.; Li, H.; Zhang, R.; Liu, L. Durability of slag and steel slag-based cementitious materials. *China Concr. Cem. Prod.* **2011**, *1*, 23–27.
78. Wang, X.; Ni, W.; Li, J.; Zhang, S.; Hitch, M.; Pascual, R. Carbonation of steel slag and gypsum for building materials and associated reaction mechanisms. *Cem. Concr. Res.* **2019**, *125*, 105893. [CrossRef]
79. You, N.Q.; Shi, J.J.; Zhang, Y.M. Corrosion behaviour of low-carbon steel reinforcement in alkali-activated slag-steel slag and Portland cement-based mortars under simulated marine environment. *Corros. Sci.* **2020**, *175*, 108874. [CrossRef]
80. Shi, C.J.; Qu, B.; Provis, J.L. Recent progress in low-carbon binders. *Cem. Concr. Res.* **2019**, *122*, 227–250. [CrossRef]
81. Zhang, B.; Zhu, H.; Feng, P.; Zhang, P. A review on shrinkage-reducing methods and mechanisms of alkali-activated/geopolymer systems: Effects of chemical additives. *J. Build. Eng.* **2022**, *49*, 104056. [CrossRef]

82. Liu, S.H.; Li, Q.L.; Han, W.W. Effect of various alkalis on hydration properties of alkali-activated slag cements. *J. Therm. Anal Calorim.* **2018**, *131*, 3093–3104. [CrossRef]
83. Hou, D.H.; Li, T.; Wang, P. Molecular dynamics study on the structure and dynamics of NaCl solution transport in the nanometer channel of CASH gel. *ACS Sustain. Chem. Eng.* **2018**, *7*, 9498–9509. [CrossRef]
84. Sun, X.; Peng, X.; Zhang, G.; Wang, S.; Zeng, L. Effect of steel slag content on properties of alkali-activated steel slag and slag based grouting material. *New Build. Mater.* **2017**, *44*, 10–13.
85. Darko, K.; Branislav, Z. Effects of dosage and modulus of water glass on early hydration of alkali-slag cements. *Cem. Concr. Res.* **2002**, *32*, 1181–1188.
86. Bernal, S.A.; de Gutierrez, R.M.; Provis, J.L.; Rose, V. Effect of silicate modulus and metakaolin incorporation on the carbonation of alkali silicate-activated slags. *Cem. Concr. Res.* **2010**, *40*, 898–907. [CrossRef]

buildings

MDPI

Article

Evaluation of Cherts in Gumushane Province in Terms of Alkali Silica Reaction

Demet Demir Şahin

Mining Technology Program, Department of Mining and Mineral Extraction, Gumushane University, 29000 Gumushane, Turkey; demetsahin@gumushane.edu.tr or demetdemir2929@hotmail.com

Abstract: Alkali–silica reaction (ASR) occurs when alkali oxides coming from the cement composition in concrete come together with reactive silica and moisture coming from the aggregate. Additional maintenance and repair costs caused by the development of ASR in concrete cause the cost to increase. However, thanks to the measures taken against ASR in the early period, it contributes to the creation of sustainable and durable concrete structures. The article investigated the usability of cherts with eight different chemical compositions as aggregates in ready-mixed concrete plants in Gümüşhane. Cherts have been investigated for alkaline reactivity and are intended to be used largely as a source for concrete plants. ASR length changes of the samples prepared according to ASTM C1260 standard were determined after 3, 7, 14, and 28 days of the curing period. Microstructural examination, ultrasonic P-wave velocity, and bending and compressive strength experiments supporting ASR were carried out. The obtained test results were compared between the reference (limestone) sample and each other, as well as with the values specified in the standards. The compressive and bending strength values of the samples increase depending on their ASR. It was observed that the crack structures and types increased depending on the increase in the crack values.

Keywords: ASR; concrete; chert; reactive aggregate; limestone; expansion

Citation: Demir Şahin, D. Evaluation of Cherts in Gumushane Province in Terms of Alkali Silica Reaction. *Buildings* **2024**, *14*, 873. https://doi.org/10.3390/buildings14040873

Academic Editor: Mohamed K. Ismail

Received: 8 February 2024
Revised: 5 March 2024
Accepted: 19 March 2024
Published: 23 March 2024

1. Introduction

Concrete, a globally fundamental construction material, represents a significant investment for national economies. Approximately 60–75% of this material consists of aggregate components, with around 10–15% comprising Portland cement or cementitious materials, along with water [1–3]. In the current era characterized by advanced technologies and finite natural resources, the production of concrete is expected to yield long life, contribute to environmental and public health, exhibit resilience to external influences, require minimal maintenance, and entail low costs [4]. Attention must be paid to the selection of aggregates, a crucial component of concrete. The chemical composition of aggregates, which are rich in compounds such as reactive SiO_2, holds significance in generating detrimental reactions within concrete. Particularly, aggregates with high silica content, strong alkali components, water, and high humidity significantly contribute to the initiation of alkali–silica reaction (ASR), a crucial factor in the deterioration of concrete [5–8].

These reactions result in substantial damage to both concrete and mortar, causing significant structural impairment [9–11]. The resolution of this damage is intricate and often demands substantial financial resources [12–15]. ASR manifests as a chemical reaction between the reactive silica in aggregates and the hydroxyl ions in the concrete pore solution (Equation (1)).

$$SiO_2 + 2NaOH + H_2O \rightarrow Na_2SiO_3 + 2H_2O \quad (1)$$

The alkali–silica reaction (ASR) converts the siloxane (Si-O-Si) bonds in the aggregate composition to silanol (Si-OH) bonds, primarily influenced by the elevated pH in the pore solution. Aggregates with a high water content in their porous structure, especially those containing opal, exhibit an abundance of silanol bonds. These bonds possess an acidic nature and react with the alkaline forms present in the pore solution. Silanol groups, upon a second attack by hydroxyl ions, transform into the SiO-form and disperse in water. Positively charged ions, such as sodium, potassium, and calcium, attract and encapsulate the negatively charged SiO-gel, leading to the formation of Equation (2) [16,17].

$$H_{0.38}\,SiO_{2.19} + 0.38NaOH = Na_{0.38}\,SiO_{2.19} + 0.38H_2O \tag{2}$$

Following the reaction, the siloxane bridge undergoes damage under the influence of hydroxyl ions, transforming into the SiO-form, resulting in swelling and the formation of a water-absorbing gel. Subsequently, the damaged area and its surroundings initiate the development of fine hairline cracks [18]. Chemical passivity characterizes the (SiO_2) in aggregates, particularly in quartz form, which are structurally arranged as (≡Si-O-Si≡). However, the irregularities on the surface of crystalline silica, their water-absorbing capability, and amorphous structure [≡Si-OH]) enhance their tendency to dissolve [19]. Consequently, silica particles exhibit a high susceptibility to dissolution. They first react with hydroxyl ions through (≡Si-OH) and, subsequently, through the neutralization of (≡Si-O-Si≡), as depicted in Equations (3) and (4).

$$\text{Si-O-Si} + H_2O = \text{Si-OH} + \text{OH-Si} \tag{3}$$

$$\equiv\text{Si-OH} + OH^- = \text{Si-O}^- + H_2O$$

$$\equiv\text{Si-O-Si}\equiv + 2OH^- = 2\text{Si-O}^- + H_2O \tag{4}$$

Sodium hydroxide (NaOH) and potassium hydroxide (KOH) from the pore solution of concrete, calcium hydroxide ($Ca(OH)_2$) produced during cement hydration, and alkali–silicate solution and gel (depending on humidity levels) begin to react. The reaction between ≡Si-O-Si≡ and hydroxyl ions intensifies further (Equations (5)–(7)) [20].

$$\equiv\text{Si-O-Si}\equiv + OH^- + R^+ \text{ (Na or K)} = \equiv\text{Si-O-Na} + \text{H-O-Si}\equiv \tag{5}$$

$$\equiv\text{Si-OH} + OH^- + Na^+ = \equiv\text{Si-O-Na} + H_2O \tag{6}$$

$$\equiv\text{Si-O-Na} + nH_2O = \equiv\text{Si-O}^- (H_2O)n + Na^+ \tag{7}$$

Following the aforementioned reactions, the alkali–silica gel penetrates the cement paste within the aggregates in concrete, initiating a reaction with it. The presence of moisture in the environment leads to expansion, stresses on the interface of the cement paste, expansion, and the formation of cracks. Reactive amorphous and weak crystal structures in natural aggregates and hydroxyl ions, along with factors such as Portland cement, aggregate particle size, admixtures, and the concrete pore solution, induce chemical reactions affecting the durability of concrete.

Concrete, being an integral element of structures, signifies strength and durability when it exhibits high-quality properties. Changes in such properties of concrete result from interactions between aggregates and cement paste, causing detrimental effects on concrete [21]. Alkali–silica reaction (ASR) is a significant phenomenon. This reaction, through a series of alterations in concrete, reduces the strength of the cement paste, leading to expansion and crack formation in concrete, thereby diminishing its strength and resistance to freezing [19,22,23]. Factors triggering alkali–silica reaction (ASR) include the type of aggregate and binder, the amorphous or crystalline state of reactive aggregates, the quantity and distribution of reactive aggregates, environmental conditions surrounding the aggregates, and various external stimulating factors, all contributing to destructive and lasting damages in concrete [24,25].

As alkali–silica reaction (ASR) induces expansion in concrete, it leads to structural damage [26,27].

The reactivity of the silica component in the composition of aggregates depends on the mineral structure. The silica component can exhibit different textures and compositions in each rock. This is primarily due to the cooling rate during the formation of rocks, which can result in either an amorphous or crystalline structure. The solubility of the silica component is higher in alkaline or acidic environments, while its solubility is lower in a neutral pH condition [28,29]. The reactivity of silica-containing aggregates is determined by the free energy of quartz. The free energy determines the solubility of quartz based on defects in the cage structure and the degree of crystallization [30]. Studies on the amorphous and crystalline structure of quartz have determined which structure is more reactive. Crystalline quartz is less soluble than amorphous quartz in a corrosive environment with a high pH ratio. This is because the amorphous structure of the aggregate, with its cracks and irregularities, allows the ions formed by ASR to penetrate the aggregate more easily than in the crystalline structure. The quality and geological characteristics of silica within crystalline quartz, such as strained quartz during formation, can contribute to its reactive nature [31]. Minerals formed rapidly due to quick cooling, such as cristobalite and tridymite, may exhibit reactive properties at normal temperatures [27,32]. Additionally, certain rocks with reactive silica include chert, some volcanic glasses, certain clays, phyllites, metamorphic breccias, as well as breccias, phyllites, schists, gneisses, gneiss granite, vein quartz, quartzite, and high-metamorphic rocks containing sandstone. However, it is not possible to definitively state that containing reactive silica will lead to ASR formation because factors such as fine-grained silica increasing contact surface with alkalis contribute to increased reactivity [18,33].

The degree of porosity of aggregate particles also plays a crucial role in the rate of ASR expansion. A higher degree of porosity allows moisture to easily penetrate into the aggregate and concrete and facilitates the easy dispersion of alkali ions [27,34].

The particle size of aggregates has an impact on reactivity. Fine-grained aggregates have silica minerals with less structural order and are more prone to instability compared to coarse-grained aggregates [32]. Particularly, when the particle size is below the given limit (~0.02–0.07 mm), it leads to less expansion, reducing the rate and amount of expansion. Coarse-grained aggregates exhibit a high expansion rate in prolonged expansion, while fine-grained reactive aggregates have a higher reaction rate [35–38]. Researchers have varying views on the influence of aggregate particle size on ASR formation. Some suggest that a fine aggregate size has a mitigating effect on ASR [23,39]. Coarse aggregates are more sensitive to ASR compared to fine aggregates. This is because fine-grained aggregates participate in pozzolanic reactions early on, converting alkali–silica gel into calcium silicate hydrate, reducing the alkali level in the solution, and promoting pozzolanic reaction [40]. The use of aggregates with particle sizes between 1 and 5 mm ensures maximum expansion [41]. It has been observed that coarse-grained aggregates cause less expansion compared to fine-grained aggregates, and reactive fine aggregates result in slow and prolonged expansion [42]. Fine reactive aggregates initiate expansion in the early stages and maintain it, while coarse reactive aggregates cause slow and prolonged expansion. As the particle size of silica-containing aggregate decreases below 0.15–10 mm, the ASR expansion increases, the cement/aggregate ratio for maximum expansion decreases, and the expansion progresses slowly [43].

The morphological structure of aggregates also influences ASR. Aggregates become angular when crushed and reduced in size. Studies have explored the impact of particle shape on ASR. Aggregates with the same type of reactive round particle structure exhibit different ASR development characteristics compared to aggregates prepared by crushing for the ASTM C1260 accelerated mortar bar test [10]. The expansion characteristics obtained by using natural and crushed aggregates of medium size at a 25% ratio differ.

It has been found that reactive particle size is more effective in crushed aggregates small and large angular particles are not effective, but medium-sized angular particles influence expansion [44].

It has been emphasized in various studies that the chemical composition alone of aggregates is not sufficient as a factor for alkali-silica reaction (ASR) formation. The reactive silica and alkali content of aggregates, particle size, particle structure, and porosity status can alter the impact value. Aggregates with amorphous, glassy, microporous structures and a large surface area, along with numerous cage defects, have a higher potential for initiating reactions (Farny and Kosmatka, 1997) [32]. Specifically, the fine-grained nature of aggregates, a large surface area, and a dense structure of cracks and pores contribute to increased reactivity [45].

- -Alkali + Reactive Silica → Alkaline Silica Gel Products
- -Alkaline Silica Gel Products + Humidity → Expansion

The presence of alkali–silica reaction (ASR) gel does not directly cause deterioration in concrete. However, high moisture or humidity in the surrounding environment leads to expansion and an increase in internal pressure. If the resulting stresses gradually exceed the tensile stresses, it can lead to the formation of cracks and, ultimately, deterioration. The ASR formation mechanism occurs as described below (Figure 1) [29,46–49].

The gel formed after the reaction does not necessarily cause significant damage to concrete every time. For the gel to cause damage, the relative humidity must be 80% or higher, leading to the spread of alkali ions, the formation of gel in reaction zones, swelling due to the absorption of water by the formed gel, and an increase in internal stresses. However, if there is an appropriate level of humidity in the concrete throughout its service life, ASR may not pose a serious threat. To prevent ASR in concrete, strategies such as low water-to-cement ratio, additional cement, mineral additives, or reducing permeability through different methods can be employed. This would hinder the spread of alkalis to the reaction zones by limiting access [50]. Additionally, the design of concrete with reactive silica-containing aggregates should ensure their incorporation into the concrete composition in an appropriate manner. This not only makes more effective use of existing resources but also significantly reduces the need for raw material exploration [51].

1. Stage

The reactive silica-containing aggregate in the concrete and the hydroxyl (OH^-) ion in the void solution of the concrete first react and then react with the alkaline components (Na^+, K^+, Ca^{++}) in the cement.

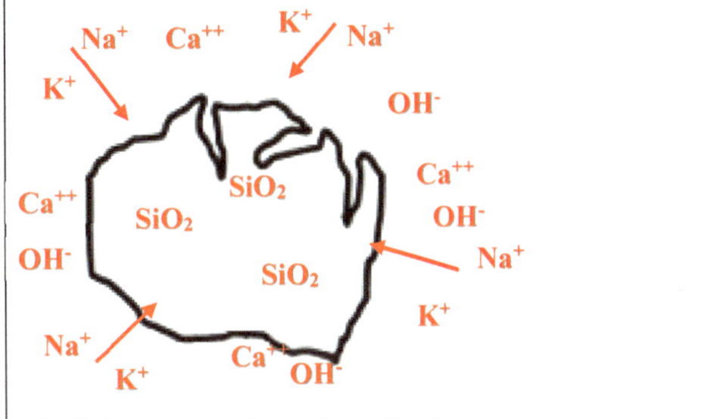

Alkali in cement and reactive silica in aggregate

2. Stage

After the reaction, a gel containing Si, Ca, Na, K and OH ions forms around and inside the aggregate.

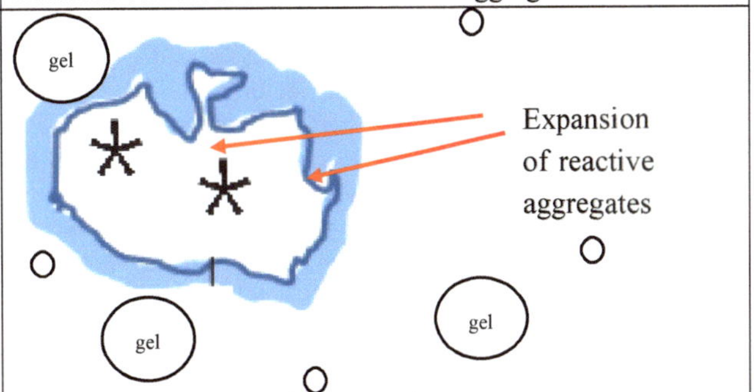

Alkaline silica gel causing expansion

3. Stage

The resulting gel causes expansion in volume due to water absorption. It causes cracks in aggregate and cake. Cracks exceeding the tensile strength cause damage.

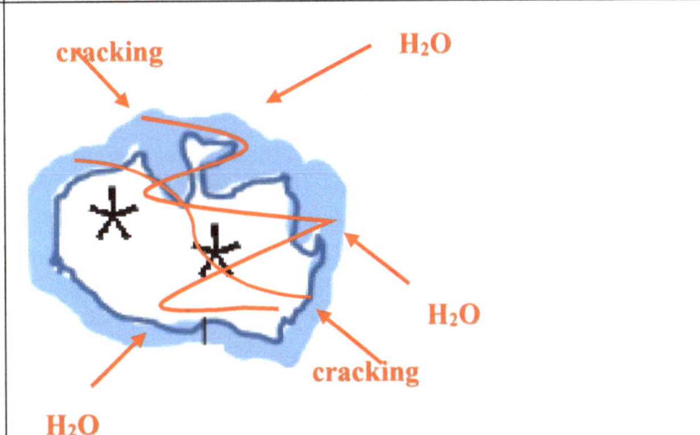

4. Stage

The pressure formed after the gel swells by absorbing water, being greater than the tensile pressure, causes expansion and formation of cracks of different sizes in the concrete.

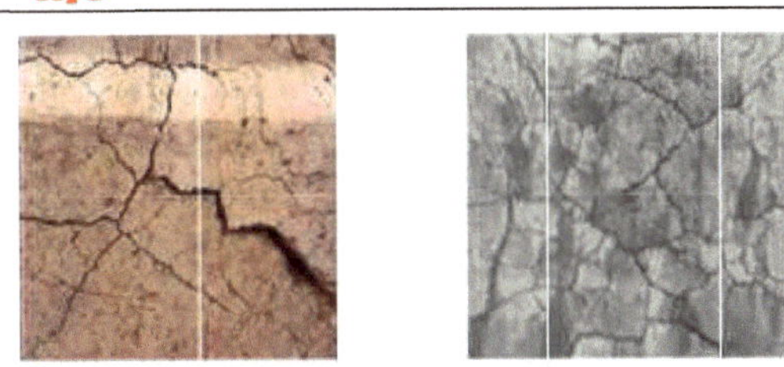

Map-like crack form in concrete after expansion

Figure 1. ASR formation mechanisms and phases of reactive aggregate in concrete composition [40,52–56].

Cherts are sedimentary rocks with a cryptocrystalline and microcrystalline structure that can lead to an alkali–silica reaction (ASR) when appropriate conditions are present in concrete. Cherts, when used as aggregates in concrete, act as known activators, causing ASR. This gel absorbs water and generates expansive forces that can lead to cracking, thus hindering the hardening process. However, due to variations in chert composition, it can exhibit different characteristic features regarding its potential to induce reactions [57]. The dissolved silica, silica content, mineralogy, chemical factors, thermodynamic properties, specific surface area, particle size, voids, cracks within particles, and crystallinity in chert composition determine its reactivity, indicating its suitability for use in concrete [58]. If chert has a weak crystalline silica structure, it tends to react with hydroxyl ions, leading to aggregate dissolution. This situation can result in the formation of alkali–silica gel, causing expansion and various durability issues in concrete [59]. The presence of a fine crystal structure in chert increases the reaction surface and enhances its reactivity [60]. Different researchers hold different views on the reactivity of cherts. Nishiyama et al. [61] stated in their study that reactive cherts have low crystallization indices. Jones (1989) [62] argued in his study that the mineralogy of cherts has a more significant impact on their reactivity than other factors.

Chert is found in limestone layers on the earth's surface or as gravel in riverbeds. This natural occurrence indicates its suitability for use as concrete aggregate. Therefore, reactive aggregates like cherts and opals do not exhibit an increasing expansion tendency. If expansion occurs, it reaches a maximum value and then decreases. The point of maximum expansion for reactive aggregates is referred to as "pessimum", and for cherts, this value is less than 10% and greater than 60% [57]. According to ASTM C33, the required chert content for ASR to occur is accepted as 3% and 8% for fine and coarse aggregates, respectively [63]. McNally and [60] considered 1% chert content as a high-expansion value in accelerated mortar bar tests. However, Gogte (1973) [64] mentioned that 1% chert content is considered high expansion, but it does not pose a problem for use unless the alkali content is high. Swamy (1992) [25] suggested in his study that concrete with more than 5% nonreactive aggregate or more than 60% chert is considered to have a low probability of being reactive.

In this study, chert samples with eight different chemical compositions were collected from various regions within the borders of Gumushane province, and for the comparison of results, limestone aggregate, which is accepted as a reference, was used. The cherts that are widespread in the region were evaluated for their potential to be an alternative concrete aggregate if used in concrete plants, which was determined using the alkali–silica reaction (ASR) elongation values. In this study, the ASTM C1260 Standard Test Method for Potential Alkali Reactivity of Aggregates (mortar-bar method) was employed to determine the alkali–silica reactivity of eight aggregates from different origins [10,29]. Sometimes, even if the aggregate is not reactive, different components may become reactive in a high alkali environment, leading to the occurrence of ASR [65].

This study aimed to identify and take preventive measures against alkali–silica reaction before it occurs in concrete, eliminating later high costs and unavoidable damages. The aggregates used in this study consisted of chert samples obtained from different routes, abundant in the Gumushane region, each having distinct chemical, mineralogical, and morphological structures. Eight chert samples were evaluated for ASR. Concrete samples, conforming to ASTM C 1260 standards, were prepared, and ASR development levels were determined after curing periods of 3, 7, 14, and 28 days. These values were then compared with the reference prepared with only limestone and the limit values specified in the standard. Additionally, chemical reactivity, particle size, water usage, pore structure of the aggregate, specific surface area determination, and concrete's compressive, tensile, and ultrasonic pulse velocity tests were conducted. The results were compared based on numerous parametric data and supported by studies available in the literature.

2. Materials and Methods

The study employed CEM I 42.5 R cement, nonreactive limestone for reference sample production, and chert samples with different chemical compositions from the Gumushane region, named TD, T, AK, GD, AC, GC, BC, and TK, as well as tap water. The origin locations of the aggregates used are provided in Figure 2 and field image Figure 3.

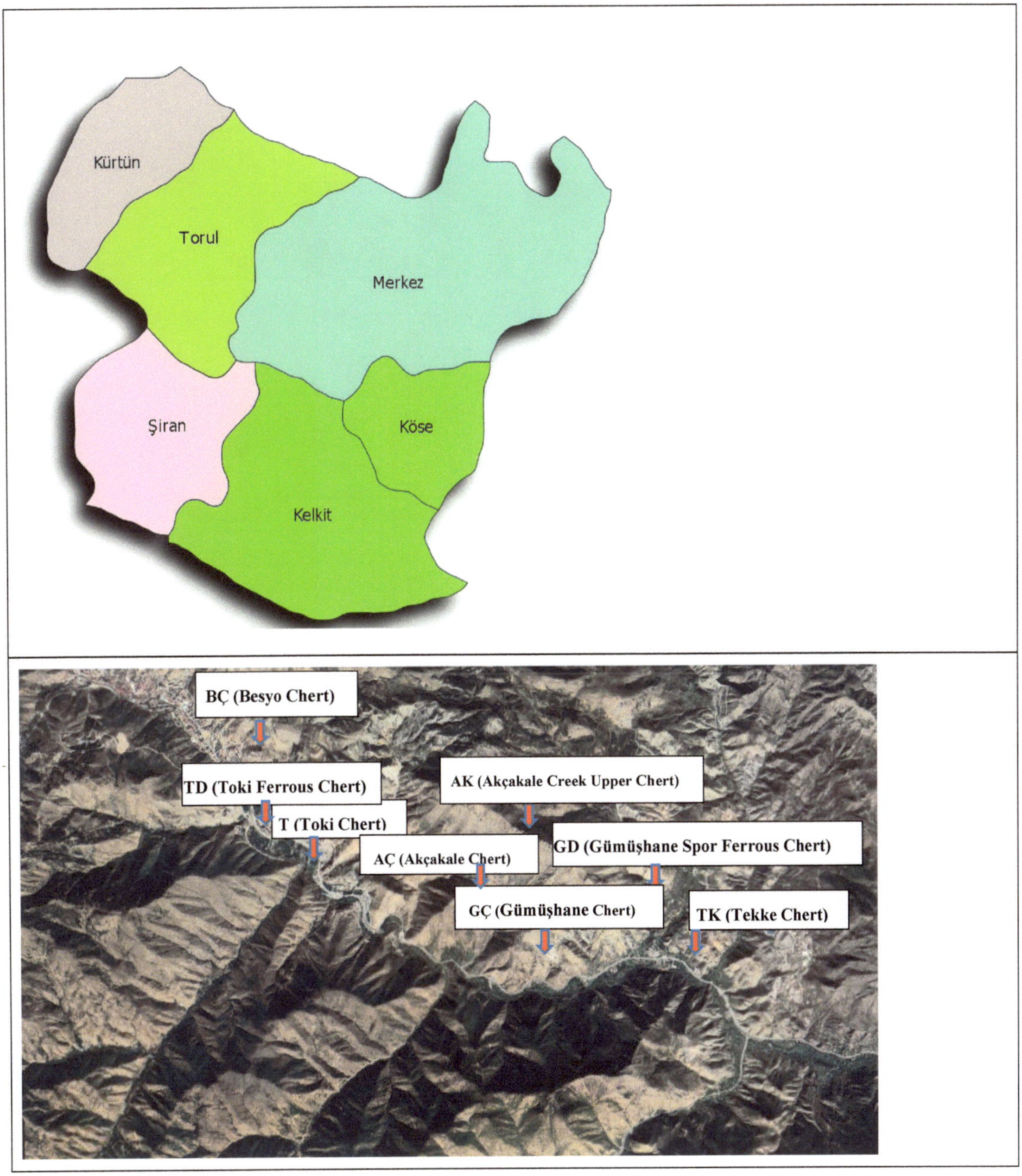

Figure 2. Where the chert is supplied.

Figure 3. Microscopic field images taken from areas where chert was collected.

2.1. Materials Used in Mortar Mix

The cement used in the mortar mixtures prepared for ASR was sourced from the Askale Cement Factory, which operates within the borders of Gumushane Province (Table 1).

Table 1. Different type properties of CEM I 42.5 R cement.

Chemical Composition of Cement (%)		Physical Parameters of Cement	
Al_2O_3	5.70	Specific Gravity (g/cm^3)	3.05
Fe_2O_3	2.99	Specific Surface Area (cm^2/g)	4140
SiO_2	19.60	Fineness value over 45 µm sieve (%)	8.50
CaO	60.39	The setting time begins (hour-min)	2 s–30 min
MgO	1.90	The setting time end (hour-min)	3 s–25 min
Na_2O	0.15	water requirement (%)	% 29.5
K_2O	0.60	Volume expansion (mm)	0.5
SO_3	2.91		
CI	0.0189	Compressive Strength (Mpa)	
s.CaO	0.34	2 days	24
Loss of ignition	7.15		
Unmeasured	0.65	28 days	52
Additive	17.87		
Total	100		

Chemical compositions of the reference limestone and different chert samples, including Toki Ferrous chert (TD), Akçakale Creek Upper Chert (AK), Toki Chert (T), Tekke Chert (TK), Gumushane Sport Ferrous Chert (GD), Akcakale Chert (AC), Besyo Chert (BC), and Gumushane Chert (GC), are provided in Table 2, while their physical properties are presented in Table 3.

Table 2. Chemical compositions and proportions of cherts.

Different Cherts	Chemical Composition of Cherts (%)								
	SiO_2	Fe_2O_3	Al_2O_3	MgO	CaO	Na_2O	K_2O	SO_3	Loss of Ignition
Limestone	3.60	0.45	1.42	19.18	29.29	0.00	0.37	0.00	42.60
TD	81.88	4.21	2.94	0.99	2.40	0.00	0.32	0.04	2.95
AK	90.54	1.88	2.41	0.09	0.66	0.10	0.42	0.00	1.38
T	83.79	0.70	0.90	1.53	3.64	0.00	0.12	0.00	5.50
TK	68.71	2.46	1.74	2.32	9.96	0.11	0.22	0.11	9.51
GD	89.19	3.22	0.37	0.00	0.30	0.00	0.00	0.00	0.16
AÇ	90.67	1.83	2.44	0.04	0.64	0.08	0.40	0.02	1.35
BÇ	54.38	0.74	0.52	6.69	14.30	0.01	0.04	0.02	10.00
GÇ	54.34	4.71	2.63	5.00	14.34	0.14	0.52	0.01	18.12

Toki Ferrous Chert (TD), Akçakale Creek Upper Chert (AK), Toki Chert (T), Tekke Chert (TK), Gumushane Sport Ferrous Chert (GD), Akçakale Chert (AÇ), Besyo Chert (BÇ), Gumushane Chert (GÇ).

Table 3. Physical properties of cherts.

Different Cherts	Fineness (%)				Specific Gravity (g/cm^3)	Specific Surface Area (cm^2/g)	Humidity Condidation (%)
	32 µm	45 µm	90 µm	200 µm			
Limestone	71.4	59.8	42.4	25.4	2.77	952	0.14
TD	58.9	23.2	3.2	0.004	2.63	4687	0.38
AK	71.5	63.9	38.6	1.4	2.64	1280	0.14
T	76.6	69.0	42.8	1.9	2.57	919	0.22
TK	69.8	61.7	38.7	1.9	2.62	1380	0.36
GD	81.7	77.6	58.4	38.2	2.61	967	0.05
AÇ	73.8	66.7	41.8	3.2	2.61	1070	0.39
BÇ	76.2	67.7	39.7	2.2	2.67	889	0.17
GÇ	76.4	68.2	48.8	5.91	2.69	1108	0.22

Toki Ferrous Chert (TD), Akçakale Creek Upper Chert (AK), Toki Chert (T), Tekke Chert (TK), Gumushane Sport Ferrous Chert (GD), Akçakale Chert (AÇ), Besyo Chert (BÇ), Gumushane Chert (GÇ).

In the study, mixture ratios for the reference sample produced only with natural limestone and mortar samples prepared according to ASTM C1260 standards with different compositions of chert samples were determined [10]. In the prepared mixture ratios, the same number of materials was used in samples produced using only limestone aggregate with reference and 8 different chert samples. Samples were prepared using 10% by weight (99 g) of No. 4 (4.75–2.36 mm) sieve, 25% by weight (247.5 g) of No. 8 (2.36–1.18 mm), No. 16 (1.18 mm–600 µm), and No. 30 (600–300 µm) sieves, and 15% by weight (148.5 g) of No. 50 (300–150 µm) sieve, according to the gradation specified in ASTM C 1260 standard. The water (206.8 g) to cement (440 g) ratio used for ASR was 0.47, and mortar mixtures were prepared. The produced mortars were placed in molds 25 × 25 × 285 mm in size and cured for 24 h at 20 ± 1 °C temperature and 50–60% relative humidity. After 24 h, the mortars were removed from the molds, and their initial lengths were measured. The samples were soaked in an 80 °C water bath for 1 day after mold removal, and their dimensions were measured. Subsequently, the samples were soaked in a 1N NaOH solution (900 mL, 40 g sodium hydroxide in distilled water) at 80 °C for 3, 7, 14, and 28 days, and periodic measurements were taken during this period.

2.2. Experiments Conducted on Materials

To determine the post-ASR elongation values of cherts with different characteristics, accelerated mortar-bar tests in accordance with ASTM C 1260 [10] standard, ultrasonic P-wave velocity measurements, and compression and flexural strength tests were conducted. Additionally, thin-section tests were performed for microstructure examinations (Figure 4).

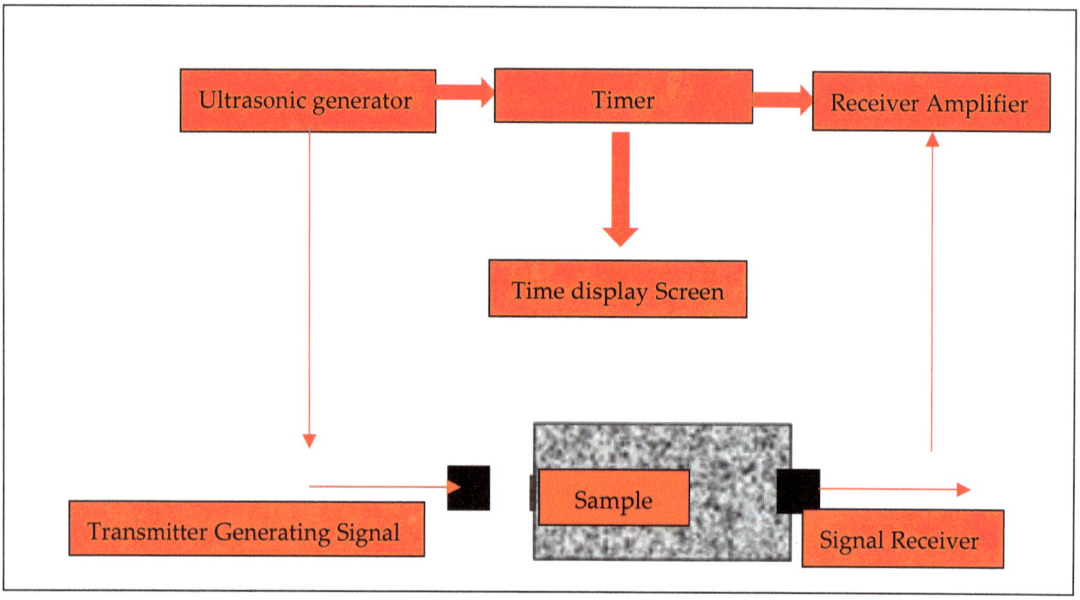

Figure 4. Ultrasonic P-wave velocity measurement setup.

The samples prepared in accordance with ASTM C 1260 standard include aggregate particle sizes from No. 4 sieve (4.75–2.36 mm) at 10% (99 g), No. 8 sieve (2.36–1.18 mm), No. 16 sieve (1.18 mm–600 μm), and No. 30 sieve (600–300 μm) at 25% (247.5 g), and No. 50 sieve (300–150 μm) at 15% (148.5 g). Prismatic samples with dimensions of 25 × 25 × 285 mm were prepared from these samples. Toki Ferrous Chert (TD), Akçakale Creek Upper Chert (AK), Toki Chert (T), Tekke Chert (TK), Gumushane Sport Ferrous Chert (GD), Akçakale Chert (AÇ), Besyo Chert (BÇ), Gumushane Chert (GÇ) chert samples used in the sample preparation stage were obtained using a jaw crusher. Three prism samples representing each material were prepared for measuring ASR elongation values after 3, 7, 14, and 28 days of curing. For comparison of the results, a reference sample was prepared using only limestone aggregate. The prepared prismatic concrete samples were kept in an oven at 80 ± 2.0 °C for 24 h in distilled water to measure the initial elongation values. Then, to determine the ASR elongation values of the samples at the specified curing periods, they were placed in a previously prepared ASR tank with a solution. The solution in the tank was prepared by adding 40 g of sodium hydroxide to every 900 mL of distilled water until the tank was filled, and the temperature indicator was set to 80 °C. After 3, 7, 14, and 28 days of curing, the samples were placed in the solution tank to measure the ASR elongation values. At the end of the curing periods, unit elongation values were determined using an ASR extensometer. The obtained elongation values were used to calculate unit elongation values as a percentage (Equation (8)).

$$\% \text{ L} = (\Delta \text{LL}) \times 100 \tag{8}$$

In Equation (8),

% L = Percentage change in height;
ΔL = For example, change in height (mm);
L = Indicates the initial length of the sample (mm).

After measuring the ASR elongation percentage results following the test, it was determined that when the elongation limit value after the 14-day curing period specified in ASTM C 1260 standard is ≤0.10%, the aggregate used is nonreactive and can be used as a harmless aggregate in terms of ASR in concrete. However, if the ASR elongation value is between 0.10% and 0.20%, it is considered suspicious in terms of ASR development, and when the limit value is ≥0.20%, it indicates that the aggregate will create reactivity in terms of ASR (ASTM C 1260, 2014).

2.2.1. Microstructural Examination of Chert Samples with Different Compositions for ASR Development

After measuring the ASR elongation values of concrete samples prepared with cherts of different compositions and properties following the 28-day curing period, the mechanism and structural characteristics of ASR formation were determined under a petrographic microscope.

2.2.2. Examination of Mechanical Properties after ASR Development in Chert Samples with Different Compositions

ASR elongation values were measured for samples produced using different chert samples after curing periods of 3, 7, 14, and 28 days. Following the measured values, mechanical properties were revealed through ultrasonic P-wave velocity, flexural, and compressive strength tests.

2.2.3. Ultrasonic-P Wave Velocity Test

The ultrasonic velocity measurement on the prepared concrete samples was determined according to ASTM C597 [66]. Ultrasonic velocity measurement was applied to concrete samples with cherts, and the operating principle of the device is provided in Figure 4. The device operates on the principle of measuring the propagation speed of the wave in the medium based on the reactions received by the receiver and transmitter sensors. The ultrasonic wave velocity test was applied to 25 × 25 × 285 mm prismatic samples. The samples were placed horizontally in the device for wave velocity measurement. Each sample was divided into three sections, and measurements were made for each section. These sections were divided into upper, middle, and lower, and three measurement values were taken for each section. The average of these 3 values reflecting the sample was calculated to obtain the result. During the measurement, the device's heads were designed to be in the same line and parallel to each other in the horizontal position of the sample. The sample heads were placed in a way that did not create gaps in the sample during the measurement, and the results obtained were calculated according to Equation (9).

$$V = (S/t) \tag{9}$$

In the given equation,
V = Ultrasonic wave velocity (m/s);
S = Length (length) of the sample (m);
t = The time (s) it takes for an ultrasonic wave to travel from one surface to another, (microseconds).

2.2.4. Uniaxial Compression Strength Test

After determining the ASR length measurement values for the prepared 3 samples with dimensions of 25 × 25 × 285 mm at 3, 7, 14, and 28 days of curing, a flexural test was conducted by first splitting them in half. Three samples were divided into two equal parts, resulting in six specimens. A compressive strength test according to the TS EN

196-1 [67] standard was applied to each specimen. After placing 40 × 40 × 40 mm-sized metal crushing heads on the lower and upper surfaces of each specimen, a compressive test was conducted. The arithmetic average of the obtained results from the 6 tests was taken and the uniaxial compressive strength values were determined by dividing the measured breaking force (F) in the device by the cross-sectional area of the specimen, as given in Equation (10).

$$f_c = \frac{F}{A_C} \tag{10}$$

In the equality,

f_c = Compressive strength, MPa (N/mm^2);
F = Maximum load reached at fracture, N;
A_C = The cross-sectional area (mm^2) of the specimen on which the pressure is applied.

2.2.5. Flexural Strength Test

Flexural strength tests were conducted on the cork samples with different components subjected to ASR. The test was carried out using a concrete press with a capacity of 200 tons and a loading speed of 50 N/s according to the TS EN 196-1 [67] standard. The samples, which had been completed during periods of 3, 7, 14, and 28 days and had measured ASR elongation values, were used for the tests. The prepared prismatic specimens were placed in the flexural apparatus, and loading was initiated. Due to the effect of the applied force (P) during loading, the specimens were split in half. The obtained measurement values were used to calculate the unknowns in Equation (11), and the flexural strength values were determined.

$$R_f = \frac{1.5 \times F_f \times I}{b^3} \tag{11}$$

In the equality,

R_f: Flexural strength, Newton/mm (MPa);
F_f: The force applied to the center of the prism at the moment of fracture (Newton);
b: The length of the side of the cross section of the prism (mm);
I: Distance between support cylinders (mm).

The materials used in the study, sample production stages, measurement of unit elongation, petrographic microscope images applied to the obtained data, ultrasonic P-wave velocity, and flexural and compressive strength tests are illustrated in a flowchart, which is presented in Figure 5.

Figure 5. ASR measurement, mechanical and microstructure investigations in concretes produced using cherts.

3. Results and Discussion

3.1. The ASR Elongation Values of Chert Samples with Different Chemical Compositions

In the conducted study, the elongation values for samples produced using limestone and eight different chert specimens for 3, 7, 14, and 28 days are presented in Figures 6–8. Assessments based on different limits according to ASTM C 1260 standards are made on the figures. The limit values specified in the standard are given in Table 4, and the comparison of the results is given in Table 5.

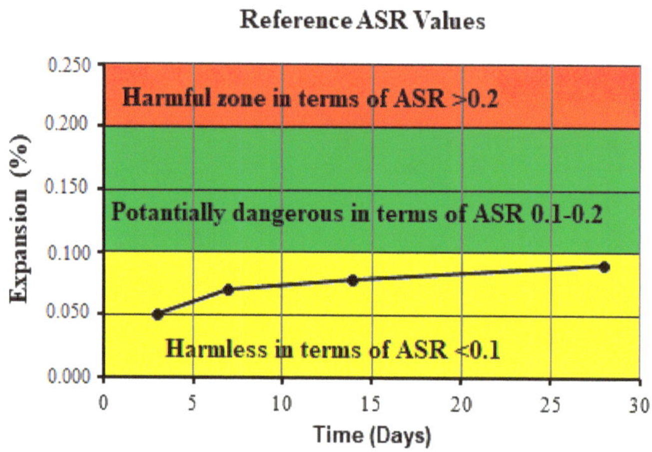

Figure 6. Reference sample ASR expansion change values.

Table 4. ASR extension limit values according to ASTM C 1260 standard.

ASR Expansion Change Values (%)	<0.1	0.1–0.2	>0.2
Determination by ASR length change method after 14 days of curing period	It is harmless in terms of ASR.	It is necessary to measure again after the 28th day, although it may be harmful. May be potentially dangerous	Harmful zone in terms of ASR

Table 5. Expansion change values of mortar bars ASR 3, 7, 14, and 28 days.

Chert Types	ASR Expansion Change Values (Day)			
	3	7	14	28
R	0.0500	0.07000	0.07800	0.09000
TD	0.0349	0.04421	0.04579	0.06211
AK	0.0754	0.09895	0.12175	0.25439
T	0.0153	0.05351	0.07947	0.19000
TK	0.0447	0.13596	0.51018	0.51667
GD	0.0065	0.18246	0.27842	0.40561
AÇ	0.2301	0.29049	0.32102	0.38067
BÇ	0.0292	0.16956	0.37921	0.41447
GÇ	0.1350	0.30151	0.33449	0.35256

Reference (R), Toki Ferrous Chert (TD), Akçakale Creek Upper Chert (AK), Toki Chert (T), Tekke Chert (TK), Gumushane Sport Ferrous Chert (GD), Akçakale Chert (AÇ), Besyo Chert (BÇ), Gumushane Chert (GÇ).

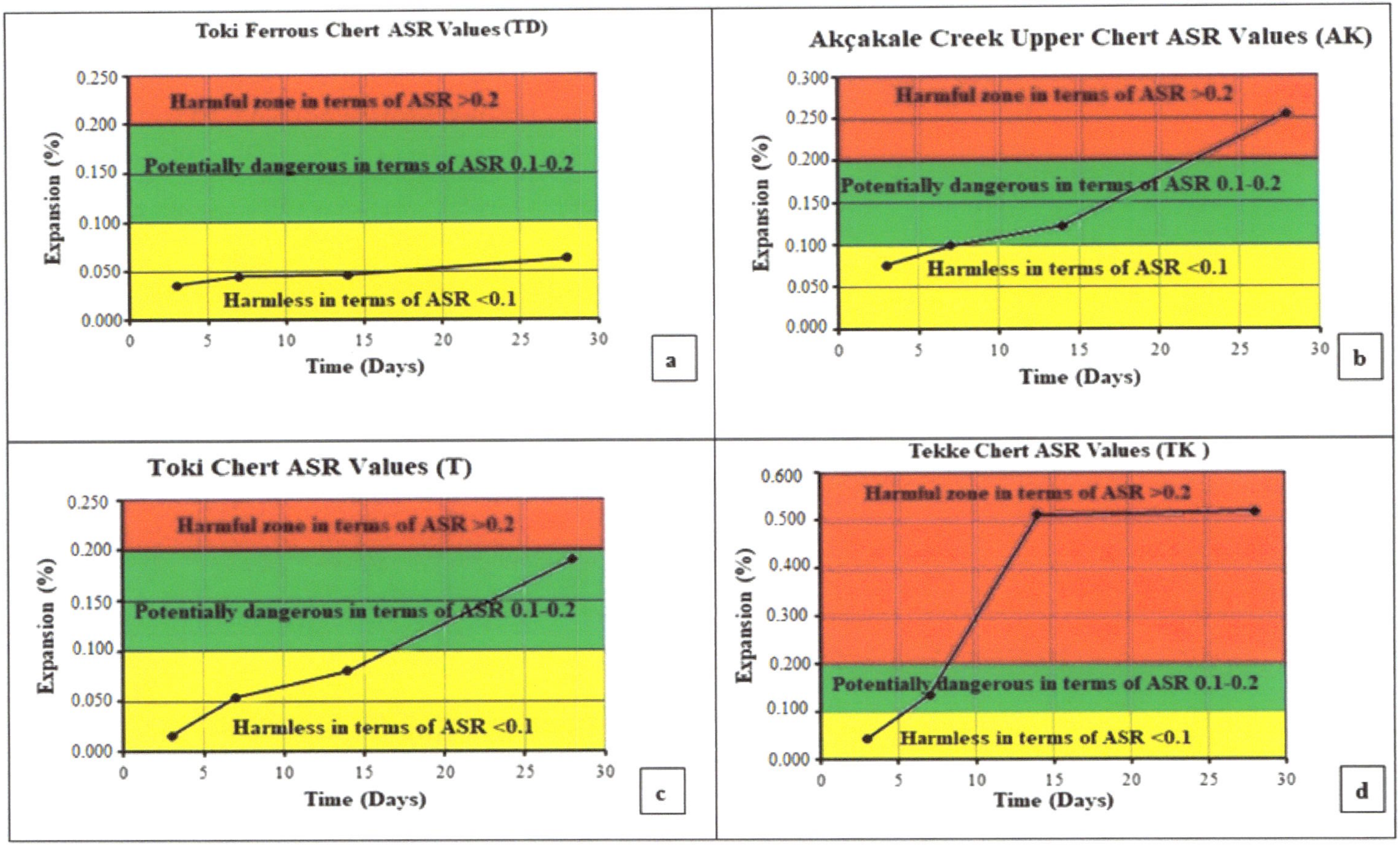

Figure 7. ASR expansion change values of different chert samples (**a**) Toki Ferrous chert (TD), (**b**) Akçakale Creek Upper chert (AK), (**c**) Toki chert (T), (**d**) Tekke chert (TK).

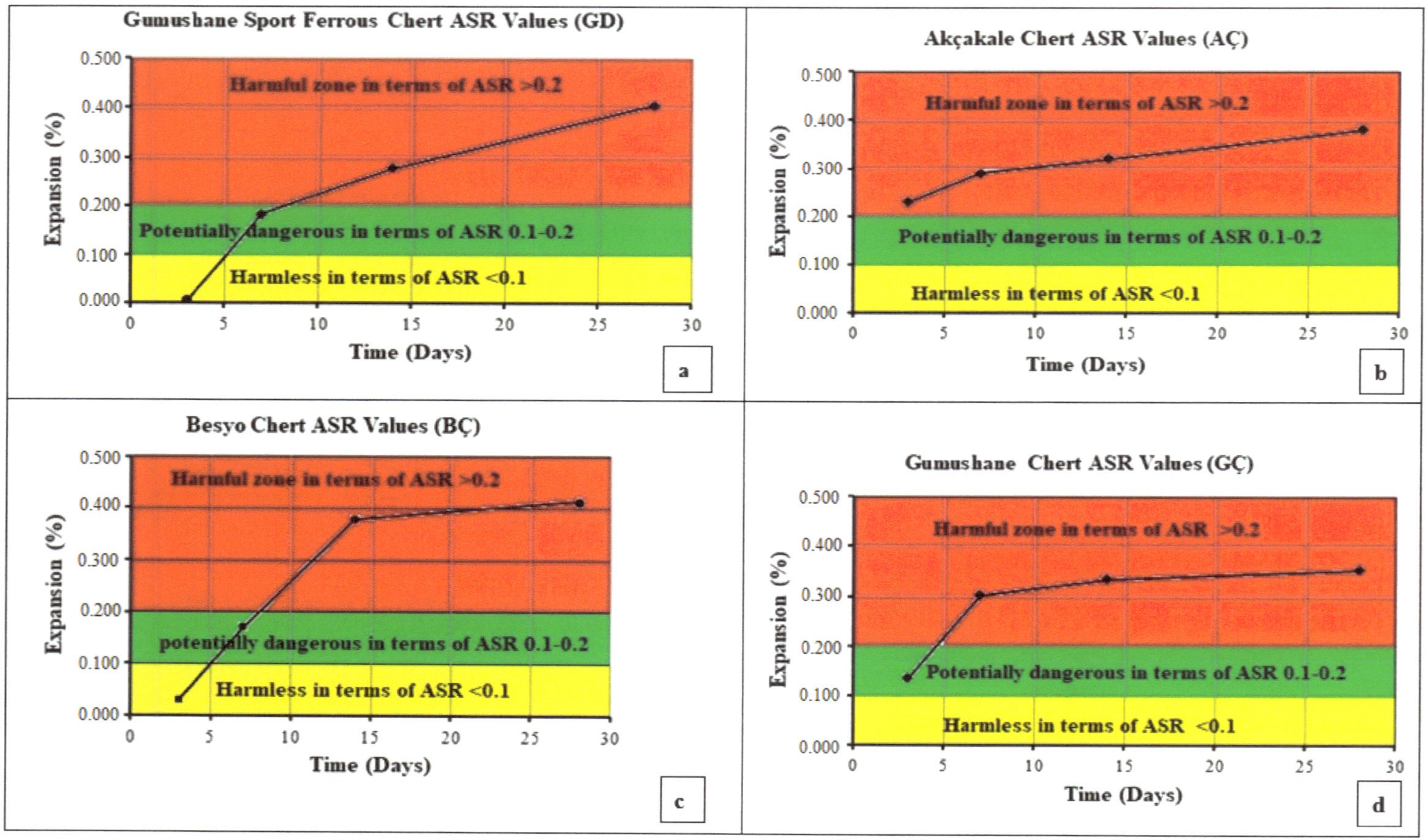

Figure 8. ASR expansion change values of different chert samples (**a**) Gumushane Sport Ferrous chert (GD), (**b**) Akcakale chert (AC), (**c**) Besyo chert (BC), (**d**) Gumushane chert (GC).

As a result of the study, the ASR elongation values of the reference and different chert samples are presented together in Figure 9. According to the obtained results, the ASR elongation values for the reference limestone aggregate increased after 3, 7, 14, and 28 days of curing. However, as the values remained below the threshold specified in ASTM C 1260 (<0.1), it was determined to be a nondamaging aggregate. Among the chert samples used in the study, the Toki Ferrous Chert (TD) and Toki Chert (T) samples also exhibited elongation values below the specified threshold, making them acceptable as nondamaging aggregates in terms of ASR development. However, other chert samples, including Akçakale Creek Upper Chert (AK), Tekke Chert (TK), Gumushane Sport Ferrous Chert (GD), Akcakale Chert (AC), Besyo Chert (BC), and Gumushane Chert (GC), showed elongation values above the threshold after the 14-day curing period. Following the 28-day curing period, the elongation values exceeded 0.2, indicating potential harm due to ASR development, and these aggregates were considered harmful.

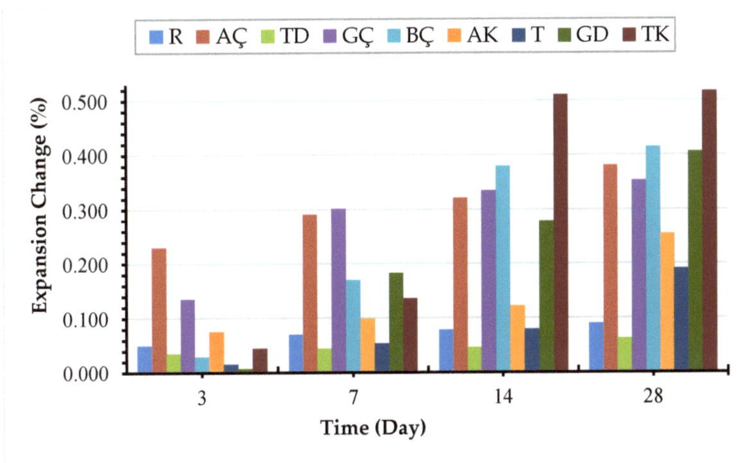

Figure 9. Results showing ASR expansion change values of reference and different chert samples.

The development of alkaline silica reaction in the AK, TK, GD, AÇ, BÇ, and GÇ chert samples used in the study was due to the presence of reactive silica, pore solution with high alkalinity and moisture in the environment. In addition, the texture difference of reactive silica, crystal structure, decreasing cooling rate during the rock formation process, silica minerals in the aggregate, amorphous or glassy structure or cryptocrystalline, crystalline and microcrystalline structure during the rock formation process, alkali content from cement and concrete, different Blaine fineness values, and the shape of the aggregates caused different elongation values of the chert samples after ASR. The reactive properties of the quartz mineral in the composition of cherts led to the development of ASR in the samples used [67]. In the studies conducted, cherts are ranked fourth in terms of forming ASR. The cherts used in this study also develop reactive silica-containing ASR. However, while ASR developed in five of eight different chert samples with reactive silica content and had height elongation values above the limit values specified in the standards, this situation did not develop in three cherts. In addition, the decrease in one of the components forming ASR causes the reaction formation to decrease or stop. However, in this study, we see that in cherts that developed ASR, the reactive silica content was sufficient until the 28-day cure period and the height growth values continued to increase.

In the studies, cherts have been accepted as a mineral showing reactivity. In addition to these, amorphous silica, opal, unstable crystalline silica, chert, chalcedony, other cryptocrystalline forms of silica, metamorphically weathered and degraded quartz, deformed quartz, semi-crystallized quartz, and pure quartz are listed according to the degree of decrease in reactivity. It is inevitable that aggregates consisting of these components will develop ASR, and chert has taken its place among the minerals with these components and gave results supporting previous studies by creating reaction. One of the components that enables the formation of alkaline silica reaction is the alkali equivalent amount used in concrete or cement expressed as sodium oxide equivalent amount. This value was calculated for all chert samples used in the study and the values obtained were evaluated in terms of ASR development (Equation (12)) (TS EN 196-2) [68].

$$(Na_2O)eşd = Na_2O + 0.658 \times K_2O \quad (12)$$

The alkali equivalent amount for the limestone aggregate and eight chert samples in cement is given in Table 6.

Table 6. Alkali equivalent amounts of cement, reference, and chert.

Sample Types	$Na_2O + 0.658 \times K_2O$ (%)	SiO_2 (%)	Blaine (cm^2/g)	Humidity, %
Çimento	0.5448	19.60	4140	0.12
R	0.24346	3.60	952	0.14
TD	0.21056	81.88	4687	0.38
AK	0.37636	90.54	1280	0.14
T	0.07896	83.79	919	0.22
TK	0.25476	68.71	1380	0.96
GD	0	89.19	967	0.05
AÇ	0.3432	90.67	1070	0.39
BÇ	0.03632	54.38	889	0.17
GÇ	0.48216	84.34	1108	0.22

Reference (R), Toki Ferrous chert (TD), Akçakale Creek Upper chert (AK), Toki chert (T), Tekke chert (TK), Gumushane Sport Ferrous chert (GD), Akçakale chert (AÇ), Besyo chert (BÇ), Gumushane chert (GÇ).

Table 6 calculates the alkali equivalent amounts causing alkali–silica reactions (ASR). The ASTM C 150 [69] standard considers a threshold value of 0.6%. The alkali values for all samples used in the study were below the specified limit in the standard, indicating nonharmful properties regarding alkali content. Cherts labeled Toki Ferrous Chert (TD) and Toki Chert (T) were not considered harmful aggregates for ASR development, unlike other cherts used in the study, including Akçakale Creek Upper Chert (AK), Tekke Chert (TK), Gumushane Sport Ferrous Chert (GD), Akcakale Chert (AC), Besyo Chert (BC), and Gumushane Chert (GC), which exhibited harmful effects in terms of ASR development due to different factors. In particular, the amount of CaO in the expanding chert aggregates showed a supportive effect on ASR formation by bringing together different components. The absence of ASR in the chert samples symbolized as TD and T in the study reflects the possibility of low values in terms of CaO composition. However, in the study, this is due to the presence of cherts with a lower value than the CaO composition in the composition of the mentioned cherts and the presence of a high reactive silica component in case of ASR formation. Wang and Gillot (1992) [70] demonstrated in their study that calcium–alkali–silica gel exhibits non-swelling characteristics. However, the Ca ions coming from the abundant CaO component in the environment replace the alkalis in ASR, allowing these components to be released into the environment. Although eight different chert bands all showed reactive properties, their ASR formation potential differed depending on their expansion values. In the studies, it has been observed that swelling reaches maximum levels when the opaline, which is accepted as reactive, is 3–5%, and if this ratio is more than 20%, no swelling is observed. The reason for the absence of swelling despite the excess of reactive minerals is firstly perceived as an unacceptable situation. This situation has arisen from the insufficient presence of all active silica in the aggregate due to the inadequate alkali oxides and the ineffectiveness of containing alkali silica gel [71]. Despite the high silica content in the aggregates referred to as TD and T in the study, the reason for the low expansion is that the alkalis present in the environment are not sufficient to react with all the excess silica. The lack of sufficient components has prevented the reaction from taking place fully [72]. The differences in ASR expansion values are also attributed to their different Blaine fineness values. Grinding finer materials during the milling process has led to smaller particles having a lower degree of structural order and unstable silica minerals compared to larger aggregates. The inverse relationship between particle size and surface area of smaller aggregates has increased the chances of attack [73]. Therefore, the increase in specific surface area of reactive aggregates has contributed to an increase in ASR expansion as particle size decreases [37,38]. The variations in ASR expansion values of mortar samples have depended on several factors. These factors include the reactivity degree of the aggregate, the chemical and mineral composition within the aggregate, crystallinity, amorphous structure, the solubility degree of amorphous silicate in the alkali pore solution, and the sensitivity of the aggregate to ASR [27]. In many studies, well-known aggregates are mostly sensitive to ASR and are classified based on their amorphous and crystalline structures. Mortar has been considered a reactive aggregate [74]. Rocks, such as opal, tridymite, cristobalite, acid volcanic glass, and basalt, that do not contain a crystal structure and exhibit an amorphous structure are irregular, internally contain microcracks, and create channels for many cage defects to easily penetrate, making them reactive [18].

The degree of porosity in aggregate particles is crucial for ASR formation. The varying composition of silica within the aggregates has led to the formation of different expansion values due to differences in texture and crystal structure. The formation of aggregates has resulted in samples with different forms and characteristics, depending on the cooling rate during the formation of silica rock, ultimately resulting in different expansion values. The silica minerals in aggregates exist in various forms such as amorphous or glassy phase (non-crystalline), cryptocrystalline, microcrystalline, and crystalline phases, depending on the decreasing cooling rate, and the formation of quartz crystals leads to variations in the expansion values of the aggregates due to the stress during crystal formation [75].

Aggregates containing strained quartz are also prone to reactivity. Cristobalite and tridymite, found in crystalline form at high temperatures and crystallizing as a result of rapid cooling, are unstable at normal temperatures. Cherts containing these crystals are reactive. Opal, an amorphous form of silica containing variable amounts of water, is highly reactive [25,32]. Additionally, it is in a direct relationship with the reactive surface area. As a result, reactive silica can be easily transferred to a larger surface area, leading to the formation of more ASR. For example, Thomas et al. (2013) [76] compared the structure of quartz (nonreactive aggregate type) with opal structure (reactive aggregate type). Quartz has a completely crystalline structure with each silicon tetrahedron bonded to oxygen ions. Each oxygen ion is bonded to two silicon ions to achieve electrical neutrality. The opal structure, despite having a crypto-crystalline and amorphous silica with each silicon tetrahedron not present, binds and dissolves oxygen ions, making them unstable [49]. In the presence of ASR, hydroxyl ions in the pore solutions penetrate silica particles, loosening the cage structure. Therefore, in crypto-crystalline or amorphous silica (such as opal), this cage structure easily breaks down with the entry of hydroxyl ions, making the aggregates susceptible to ASR. In contrast, well-crystallized silica is not sensitive to ASR attacks due to its regular cage structure and oxygen ions bonded to each silicon. Poorly crystallized, heavily cage-defective, amorphous, glassy, and microporous aggregates have shown more susceptibility to ASR formation [32].

The different Blaine fineness values of the cherts also contributed to variations in ASR elongation values becaues micro-cracks within aggregates affect the reaction surface area. The particle size of aggregates being fine or coarse has different implications for ASR formation. Some researchers have argued that a certain size of expansion occurs at maximum levels (0.07–0.85 mm), and particles larger and smaller than this size reduce expansion [77]. Mehta (1993) [41] and SHRP (2003) [78] emphasized in his study that the particle size for the maximum expansion caused by reactive aggregates is between 1 and 5 mm. The differences in reactivity of the chert aggregates used in the study could be attributed to different Blaine fineness values, as well as different alkali, silica, and calcium content. Nishibayashi and Yamura (1992) [79] and Ramyer et al. (2005) [44] determined in their research that the initial expansion of concrete occurs when only fine and reactive aggregates are used, stabilizing later. Similarly, using only reactive coarse aggregate leads to slow and prolonged expansions. The dimensional differences of aggregates have different effects on ASR. Fine aggregate size is more sensitive to ASR formation compared to coarse aggregate [32,35]. This is because smaller particles during the grinding process have more unstable minerals and lower structural defects compared to larger particles [72,78].

3.2. Microstructural Impact of Chert Samples with Different Compositions on Alkali–Silica Reaction (ASR)

For the alkali–silica reaction (ASR) to occur, certain mineralogical components must be present. The determination of these components is carried out using the ASTM C295 [9] standard, titled "Standard Test Method for Petrographic Analysis of Aggregates". In the method, samples causing ASR are examined under a petrographic microscope, and the formations are presented in Figure 10. The polarizing microscope images in Figure 10 were obtained with double nicol and objective lens magnification ratios of 4×0.10. Following the microscopic examinations, ASR formations, mineralogical compositions, and details are provided in Table 7. Particle characteristics, such as shape, size, texture, color, mineral composition, and physical conditions observed in thin sections under the petrographic microscope, are classified according to mineralogical compositions, texture, formal properties, and ASR formations expressed in the ASTM C295 [9] standard [80].

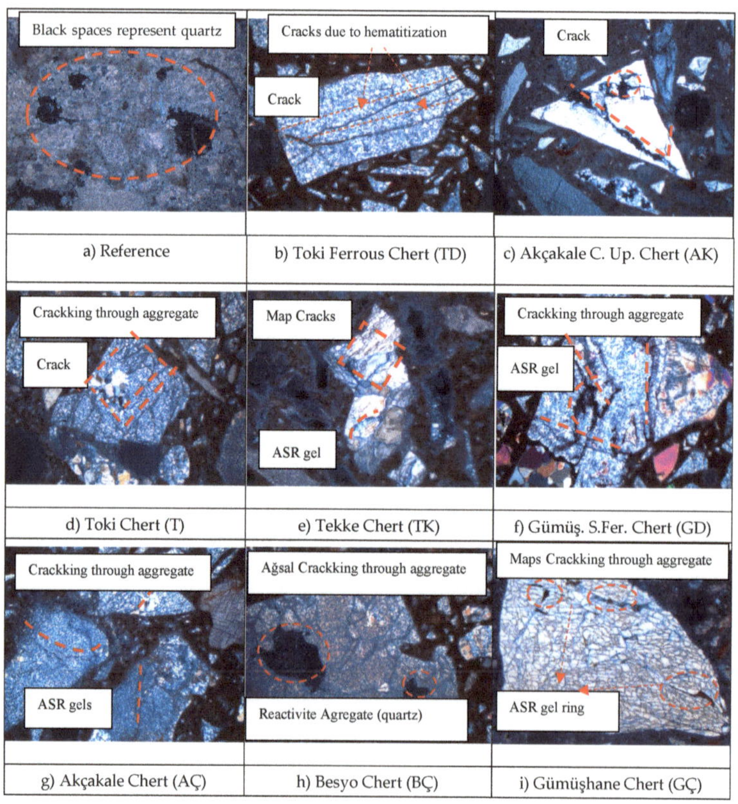

Figure 10. Alkali silica reaction thin-section images.

Table 7. Petrographic analysis results of cherts.

Samples	Description	ASTM C295 Petrographic Classification of ASR
Reference (R)	Black voids are quartz, and pink voids are sericite and clay. It does not form ASR.	CLASS I
Toki Ferrous Chert (TD)	The cracks in the aggregate are due to haematitisation from the iron mineral in the chert and do not cause ASR	CLASS I
Akçakale Creek Upper Chert (AK)	The triangular grain is composed of amorphous quartz, and the cracks on it are due to ASR.	CLASS III
Toki Chert (T)	There is secondary quartz formation and hematitisation. The cracks, which were initially due to hematitisation, were later formed as a result of ASR and manifested themselves by taking a map image.	CLASS II
Tekke Chert (TK)	Cracks due to ASR were intensively formed on the aggregate, and these showed themselves in the map appearance.	CLASS III
Gumushane Sport Ferrous Chert (GD)	There are ASR gels formed around the aggregate and map cracks surrounding its interior	CLASS III
Akçakale Chert (AÇ)	It contains chalcedony. ASR gels and network-like cracks were formed around and inside the aggregate grains.	CLASS III
Besyo Chert (BÇ)	There are two different quartz grains in the aggregate and a network structure formed due to ASR.	CLASS III
Gumushane Chert (GÇ)	The aggregate is composed of quartz grains. The quartz is dominated by intermittent ASR gel formations and network crack structure.	CLASS III

CLASS I: Far from showing alkaline reactivity, CLASS II: Alkali reactivity uncertain, CLASS III: May show alkali reactivity.

Binal (2004) [81] determined the alkali reactivities of five different reactive aggregates, namely opal nodule, chert, chalcedony nodule, andesite, and basalt, in terms of their petrographic, mechanical, and physical properties. Petrographically, different alkali–silica gel developments were identified under the microscope based on the types of reactive aggregates. The development of alkali–silica gel in these aggregates was determined with different structural images under the microscope, including halo-shaped formations in opal minerals, concentric halo formations along radial cracks, and radial fibrous alkali–silica gel images in chert minerals due to the aggregate being composed of fibrous cryptocrystalline quartz. Additionally, alkali–silica gel images, in the form of radial fibers originating from the aggregate, were observed in chalcedony minerals. The results obtained correspond with the elongation results, reflecting the correlation between petrographic analysis and ASR formation mechanisms. It should be noted that, while petrographic analysis results are considered in this study and in previous research, they are utilized as an auxiliary method, and the ASR formation mechanism is not determined solely based on petrographic analysis. According to the British Specification for Highway Works, a material is considered nondetrimental if it is not contaminated by reactive silica minerals, such as opal, tridymite, and cristobalite, and if it does not contain more than 2% by mass of chert, flint, or chalcedony. Quartz should not contain more than 30% by mass of metamorphosed quartzite. Petrographic analyses in this study were supported by the elongation results to determine minerals causing ASR.

In Figure 10, the structure of mineralogical components causing ASR, as well as the condition of the aggregate and pore structure, have significantly contributed to the initiation and propagation of the alkali–silica reaction. The porosity of the aggregate is an effective factor in ASR. A higher porosity structure has made the concrete and aggregate more sensitive to ASR, i.e., it has increased its susceptibility [27,34]. The porous structure of both concrete and aggregate with high porosity easily absorbs the moisture causing ASR, facilitating the transport of different components. The rapid spread of alkali ions in porous aggregate initiates the dissolution of the aggregate causing ASR [37]. In the study conducted by Eker et al. (2023) [82], they examined the effect of the fineness of fly ash on the ASR formation mechanism. They compared the microstructural changes of the reactive reference aggregate causing ASR with samples containing fly ash. The ASR in the reference concrete had the highest elongation value, as observed in the images obtained; there were cracks in the macrostructure, ASR gel formation inside the aggregate, and large and small cracks around or along the aggregate. The reducing effect of fly ash fineness and replacement ratio on ASR was attempted to be demonstrated through petrographic analyses. According to the obtained images, increases in the fineness of fly ash and replacement ratio led to decreases in voids and crack structures. After perographic examination, the absence of ASR in the chert samples symbolized as R, TD, and T is due to the lack of structures and minerals that will form ASR in terms of crystal structure. In addition, the absence of fractures and cracks in the cherts, as well as the mineral structure, caused products that would form ASR to enter the aggregate, and no reaction developed.

3.3. Examination of Mechanical Properties of Chert Samples in Different Compositions of Aggregates following ASR Development

Ultrasonic P-Wave Velocity

After measuring the ASR elongation values of the samples following the 28-day curing period, ultrasonic P-wave velocity and flexural and compressive strength tests were conducted. All the results of the experiments are presented in Table 8. Figures 11–13 depict the mechanical data for the samples.

Table 8. At the end of 28-day curing period, all mechanical results of the specimens.

Sample Names									Experiments
R	**AK**	**TD**	**GÇ**	**BÇ**	**AÇ**	**T**	**TK**	**GD**	
3683	3674	3685	3391	3300	3427	3782	3253	3557	Ultrasonic Wave Velocity (m/sn)
9.32	5.17	9.71	3.05	2.93	4.31	9.14	2.76	4.29	Compressive Strength (MPa)
2.77	2.44	2.83	2.24	2.18	2.33	2.84	1.98	2.40	Flexural Strength (MPa)

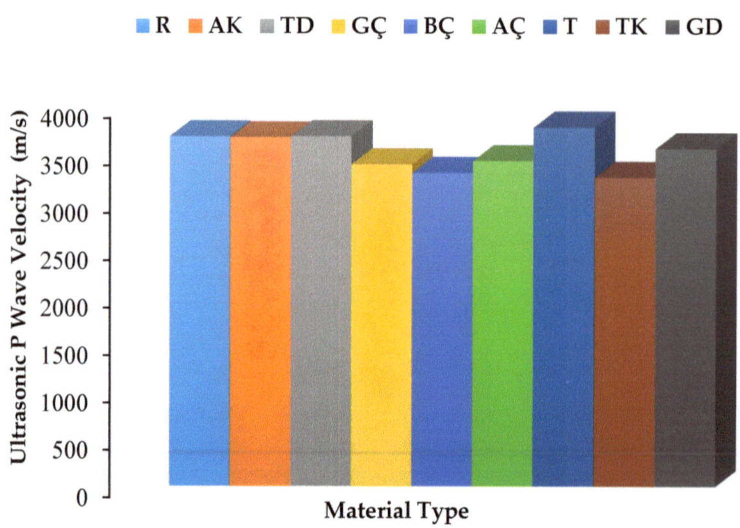

Figure 11. Ultrasonic P-wave velocity results for chert samples.

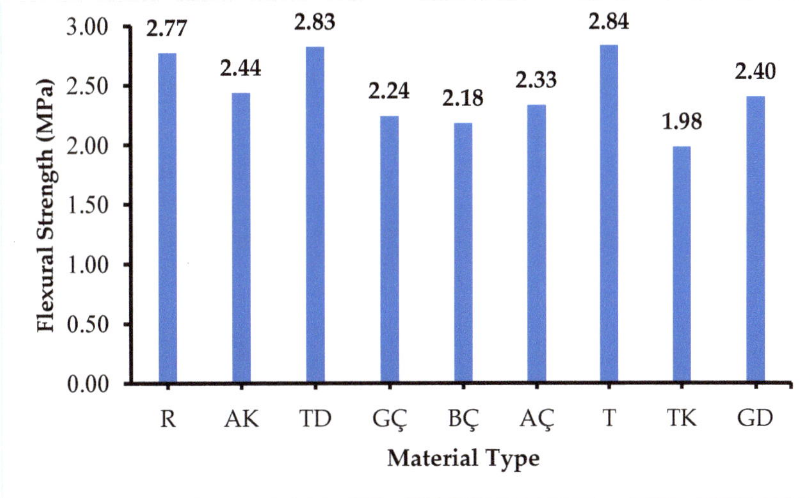

Figure 12. Flexural strength value results of chert specimens.

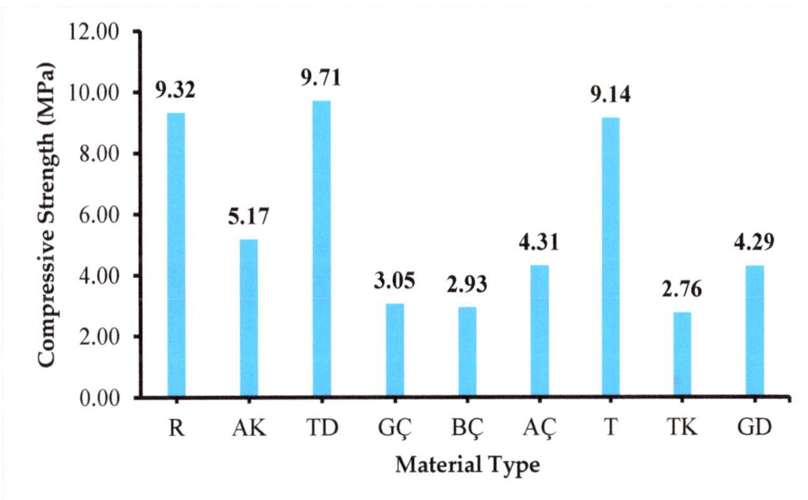

Figure 13. Compressive strength value results of chert specimens.

After ASR, the sample with the highest elongation value had the lowest ultrasonic wave velocity value, while the sample with the lowest elongation had the highest ultrasonic wave velocity value. Following a 14-day curing period, the highest ultrasonic P-wave velocity of 3685 m/s was determined for the TD mortar sample with the lowest ASR elongation value, while the highest ASR elongation value belonged to the TK sample with the lowest ultrasonic P-wave velocity of 3253 m/s (Figure 11).

The ultrasonic P-wave velocity results of the aggregates used in the study were compared in Table 9 in terms of concrete quality, as per the studies conducted by [83,84].

Table 9. Evaluation of ultrasonic P-wave velocity values in terms of concrete quality.

Ultrasonic P-Wave Velocity (m/sn)	Concrete Quality	
	Whitehurst, 1951 [82]	Uyanık et al., 2011 [83]
Excellent	>4500	>4565
Good	3650–4500	3515–4565
Medium	3050–3650	2930–3515
Poor	2000–3050	2110–2930
Very Poor	<3000	<2110

The samples represented by symbols T, TD, R, and AK, when used as aggregates in concrete, result in high-quality concrete, while concrete produced with the samples represented by symbols GD, AC, GC, BC, BK, and TK exhibit medium-quality characteristics. After ASR, there was no significant decrease observed in ultrasonic P-wave velocities. It is indicated that the changes in ultrasonic P-wave velocity values are influenced by the mineralogical compositions within the aggregate, but this influence leads to relatively lower variations compared to microcrack structures [45].

4. Flexural Strength Test Results

To determine the mechanical strength losses in mortar samples prepared using chert samples after the ASR effect, flexural and compressive tests were conducted on samples aged for 28 days under water curing, and the obtained flexural strength values are presented in Figure 12.

The flexural strength values of the samples exhibited elongation in their lengths and mechanically manifested themselves by obtaining low flexural strength values after exposure to ASR. As the elongation value increased after ASR, a decrease in flexural strength values was observed. Among the cherts symbolized as R, TD, and T, which did not induce ASR according to standards, the highest flexural strength values were obtained. For the remaining samples symbolized as AK, GD, AC, GC, BC, and TK, a decrease in elongation values corresponded to an increase in flexural strength values. The highest flexural strength value was 2.84 MPa for the sample symbolized as T, while the lowest flexural strength value was 1.98 MPa for the sample symbolized as TK. The difference between the highest and lowest flexural strength was 0.86 MPa, representing a strength loss of 30.28%. Demir (2010) [85] determined flexural strength values in mortar samples prepared with the addition of silica fume and fly ash after ASR. It was found that the flexural strength values of the silica fume were 18% better than those of the fly ash-added samples. When reactive aggregate is used in concrete, it has been concluded that it has significant effects on guiding properties such as flexural and compressive strength values. Studies have shown that after the ASR development of mortar samples, concrete flexural/tensile and elastic modulus are affected [86]. The flexural strength values obtained for the mortar samples have yielded very close results. This indicates that flexural strength values may not always change at the same rate and speed. The crack structure and formations developed in mortars after ASR, along with other mechanisms contributing to the mineral composition inside the aggregate and ASR formation, easily affect the flexural strength values [25,45,87].

Compressive Strength Test Results

After determining the ASR length measurement values of mortar samples that completed the 28-day curing period, flexural strength was measured, followed by the determination of compressive strength values. The obtained results are presented in Figure 13.

A decrease in compressive strength values was observed after ASR. The decrease was more pronounced in the sample with the highest elongation value due to ASR. The compressive strength values, from highest to lowest, were ranked as TD, R, and T for samples without ASR, and as AK, GD, AC, GC, BC, and TK for samples with ASR. The percentage loss of pressure values between samples, from highest to lowest, is given in Table 10.

Table 10. Compressive loss due to ASR between chert specimens.

Compressive Loss Values between the Four Specimens after ASR (%)							
TD-R	R-T	T-AK	AK-GD	AÇ-GD	AÇ-GÇ	GÇ-BÇ	BÇ-TK
4.02	1.93	43.44	17.02	0.46	29.24	3.94	5.81

Among the chert samples, the highest compressive strength value belonged to the TD sample, with a value of 9.71 MPa, while the lowest compressive strength value for the TK sample was calculated as 2.76 MPa. The pressure loss between the two chert samples was 6.95 MPa. A pressure loss of 71.6% was determined between the TD sample without ASR and the TK sample with the highest elongation value due to ASR. Deformation structures formed in the samples were supported by pressure and flexural strength losses. In his study, Demir (2010) [85] applied an ASR test on mortar samples prepared with the addition of silica fume and fly ash. After the experiment, it was determined that ASR occurred, elongation values increased, and compressive strength values decreased in the samples. In the samples used in this study, the development of ASR has led to cracks in the internal structure of the concrete, resulting in pressure losses.

The decrease in compressive strength values of mortar samples prepared with ASR affected by the formation of a gel with a high water retention capacity. This gel caused expansion and increased internal stresses, resulting in a loss of strength reflected in the sample results. Over time, the gel, which increased in volume by absorbing water, led to the formation of micro-cracks between the aggregate and cement paste. The continuous water absorption by the gel within the micro-crack structure further expanded the gel, enhancing and multiplying additional expansions and crack formations. Such structures in concrete or mortar have led to damage in concrete, causing reductions in mechanical properties such as compressive strength, flexural strength, and ultrasonic pulse velocity values [86–88].

5. Conclusions and Recommendations

In this study, experiments were conducted using eight different chemical composition chert aggregates from Gumushane province and one nonreactive reference aggregate (limestone). The following results have been revealed:

- The suitability of the reference sample composed of nonreactive limestone and the eight chert samples from different regions of Gumushane province for use as aggregates in concrete was evaluated in terms of ASR development. All the cherts used in the study contained high levels of reactive silica, and ASR development was not observed in all of them. In general, in the cherts that developed ASR, the fine texture of silica and the increase in the contact surface with alkali increased the ASR development.
- The chemical compositions of the cherts, which was determined using the XRF method, revealed varying amounts of reactive SiO_2 composition from high to low, resulting in the following order: AC, AK, GD, T, TD, TK, BC, GC, and R. In terms of reactive silica content, the chert called T ranks fourth, TK ranks fifth and R ranks ninth. Although reactive silica content is high in these cherts, ASR did not develop. In this case, the crystalline structure of T and TK cherts and R limestone samples is dominated by the crystalline structure of AE, while the amorphous structure is dominant in AK, GD, TK, BÇ, and SE cherts. Amorphous cherts are dominated by fractures and irregular structures, allowing ASR products to penetrate the aggregate through these irregular structures. As a result, these cherts caused the development and progression of ASR.
- Alkali equivalent amounts were calculated for the reference and eight chert samples according to ASTM C 150 standard, with values below the specified limit of 0.6%. Although the alkali content in the chert samples was below the limit value specified

in the standards, no reaction developed in the cherts expressed as TD and T, but it developed in AK, TK, GD, AÇ, BÇ, and GÇ cherts. The reason for this is the diffusion of alkali ions in the reacted cherts, gel formation in the reaction zones, and the presence of irregular internal structures that cause swelling and increase the internal stress by water absorption of the gel formed.

- The physical properties of the cherts were determined through specific surface area and sieve analysis. The highest specific surface area was found in the TD sample, with 4687 cm^2/g, while the lowest was in the BC sample with 889 cm^2/g. There was an 81.03% decrease between the highest and lowest specific surface area values. The ASR mechanisms were positively affected by specific surface area values, indicating that the TD chert sample with the highest specific surface area did not develop ASR, while the BC chert sample with the lowest specific surface area did. By comparing the ASR developments according to the specific surface area values, different situations were observed in the samples. In the TD chert sample, which has a high specific surface area, its high fineness caused less expansion, expansion rate, and amount, and ASR did not develop. The reason is that the high specific surface area value in the aggregate caused the formation of a pozzolanic reaction in the early period, converting the alkali silica gel into the calcium–silicate–hydrate formula, reducing the alkali level in the solution and causing the Pozzolanic deterioration. It was predicted that the development of alkali silica ratios in the T chert sample, which has a low specific surface area, has not yet occurred and may develop over a longer period of time. In other cherts, ASR events occurred through its textural structure and internal defects, regardless of whether the specific surface area value was high or low.
- In the conducted study, elongation values were determined according to ASTM C 1260 standard for reference and eight chert samples after curing periods of 3, 7, 14, and 28 days. Regarding ASR, samples were considered harmful aggregates if the elongation values exceeded 0.1 after the 14-day curing period and exceeded 0.2 after the 28-day curing period. According to this standard, samples denoted as R, TD, and T obtained elongation values below the threshold and were thus considered harmless aggregates in terms of ASR development. On the other hand, chert samples labeled AK, TK, GD, AC, BC, and GC were determined to exceed the specified threshold elongation values, indicating their potential for ASR development.
- After the development of ASR, thin sections were prepared to observe microstructural changes in the samples, and they were examined under a petrographic microscope. According to the obtained results, samples denoted as R, TD, and T did not exhibit ASR development, as they obtained elongation values below the threshold, and no associated crack structure was observed. On the other hand, the chert samples labeled AK, TK, GD, AC, BC, and GC showed ASR development, and under the petrographic microscope, map cracks, cracks along the aggregate, ASR gel, vesicular cracks, and fine hairline crack structures were observed. It was observed that the development of ASR in the chert samples included in the study did not depend only on the reactive silica content. Reactive silica content, specific surface area values, and the effectiveness of the grain structure were observed. In addition to these features' porosity status, AK, TK, GD, AC, BÇ and GÇ are microporous and have many lattice defects. While the potential for reaction is high, in TD and T samples, the reaction does not develop. ASR has developed in chert particles with high porosity, allowing moisture to easily penetrate and alkaline ions to disperse easily.
- Ultrasonic P-wave velocity, flexural, and compressive strength tests were conducted to determine mechanical changes in the samples after ASR development. After these tests, the quality of the concrete produced with the chert used is determined. The fact that concrete has a strong and durable feature means that a quality concrete is produced. Changes that may occur in the quality of concrete are due to the ASR formed between the aggregate and cement paste in its composition. This reaction, after a series of changes in the concrete, reduced the strength of the cement paste, causing

expansion and cracks in the concrete, and a decrease in mechanical values such as bending, pressure and ultrasonic P-wave velocity.

- Based on the ultrasonic P-wave velocity test results, it was estimated which type of concrete would be produced according to the classification expressed in studies determining concrete quality. Aggregates produced from cherts represented by T, TD, R, and AK were found to be of good quality, while concrete produced from cherts represented by GD, AC, GC, BC, BK, and TK showed moderate quality. It has been observed that ASR has not developed or has not started yet in the concretes produced with the T, TD, and AK chert samples and R (limestone) used. It has been observed that ASR started and continued in concretes produced with chert samples using GD, AC, GC, BC, BK, and TK, thus reducing the quality of the concrete.
- A flexural strength test was conducted to determine the mechanical losses of the cherts assumed to be used as aggregates in concrete after ASR. Cherts symbolized by R, TD, and T, which did not induce ASR elongation according to standards, showed the highest flexural strength values. The remaining cherts symbolized by AK, GD, AC, GC, BC, and TK showed an increase in flexural strength values as the elongation values decreased. The highest and lowest flexural strength values showed a decrease of 30.28%. Compared to the compressive strength values, the flexural strength values of the chert specimens that did not develop ASR, and the chert specimens that developed ASR did not show sharp decreases. This is because the reductions in the tensile and flexural strengths of concrete do not always occur at the same rate or rate. In these cases, it is difficult to expect a single critical expansion limit to be applicable to all structures. When considering the use of these limits, it is necessary to take into account changes in the engineering parameters of the structures, the type of aggregate, and the rate of reactivity leading to a decrease in these parameters.
- Another factor determining the mechanical strength losses of chert samples is compressive strength, which was applied to the samples. Compressive strength values, from highest to lowest, were arranged as follows: in samples in which ASR did not occur, TD, R, and T and in samples in which ASR occurred, AK, GD, AC, GC, BC, and TK. The chert sample with the highest compressive strength value belonged to the TD sample, with a value of 9.71 MPa, while the lowest compressive strength value was calculated for the TK sample at 2.76 MPa. A pressure loss of 71.6% was determined between these two samples. The use of ASR-forming cherts in concrete caused swelling and volume increase. The resulting volume increase caused tensile stresses and cracks. After the test results obtained, the lowest compressive strength value was observed in the chert sample expressed as TK. At the end of the 14- and 28-day curing periods, expansion values of 0.5% and more were obtained due to ASR formation. When chert samples with low compressive strength, such as AK, GD, AÇ, GÇ, and BÇ, which form ASR in concrete, were used, cracks were formed and these formations were observed in microscope images. This is because it was observed that cherts such as AK, GD, EEA, SE, and BÇ had expansion values higher than expansion unit deformation values such as 0.04–0.05%, which caused cracking.
- Of the eight different chert samples used in the study, five develop ASR, while three do not. It is necessary to determine which methods provide alkali silica reclamation of five different samples or which policies should be used together in the future and the method of using them as a source for regional concrete production. If this is achieved, the region will be able to produce easy-to-obtain and cost-effective concrete and meet the need for new aggregate resources in the region. To prevent ASR development of these cherts when used in new studies, the following recommendations can determine their safe use in different studies. These suggestions are as follows:

    1. Use cement with a cement alkaline component that will provide ASR formation lower than the value specified in ASTM C 150 as 0.6% Na_2Oeq. Deprive alkali that would enable ASR to form may prevent the reaction from occurring;

2. The use of cement with low alkalinity additives instead of the cement specified in the ASTM 1260 C standard may limit the formation of ASR reactions;
3. The cement to be added to ASR should be made of silicon-rich F class fly ashes and slag together or used separately;
4. Investigate the use of only blended cement and its effectiveness.

Funding: This research received no external funding.

Data Availability Statement: The original contributions presented in the study are included in the article.

Acknowledgments: I would like to thank the İbrahim AKPINAR of Gumushane University.

Conflicts of Interest: The author declares no conflicts of interest.

References

1. Naik, T.R. Sustainability of concrete construction. *Pract. Period. Struct. Des. Constr.* **2008**, *13*, 98–103. [CrossRef]
2. Kolawole, J.T.; Babafemi, A.J.; Paul, S.C.; du Plessis, A. Performance of concrete containing Nigerian electric arc furnace steel slag aggregate towards sustainable production. *Sustain. Mater. Technol.* **2020**, *25*, e00174. [CrossRef]
3. Adiguzel, D.; Bascetin, A.; Baray, Ş.A. Determination of Optimal Aggregate Blending to Prevent Alkali-Silica Reaction Using the Mixture Design Method. *J. Test. Eval. ASTM Int.* **2019**, *47*, 43–56. [CrossRef]
4. Figueira, R.B.; Sousa, R.; Coelho, L.; Azenha, M.; de Almeida, J.M.; Jorge, P.A.S.; Silva, C.J.R. Alkali-silica reaction in concrete: Mechanisms, mitigation and test Methods. *Constr. Build. Mater.* **2019**, *222*, 903–931. [CrossRef]
5. Davraz, M.; Gündüz, L. Reduction of alkali silica reaction risk in concrete by natural (micronised) amorphous silica. *Constr. Build Mater.* **2008**, *22*, 1093–1099. [CrossRef]
6. Forster, S.W.; Boone, R.L.; Hammer, M.S.; Lamond, J.F.; Lane, D.S.; Miller, R.E.; Parker, S.E.; Pergalsky, A.; Pierce, J.S.; Robert, M.Q.; et al. *State-of-the-Art Report on Alkali-Aggregate Reactivity Reported by ACI Committee*; ACI Committee: Indianapolis, IN, USA, 1998; ACI 221.1R-98; pp. 1–31.
7. Latifee, E.R. State-of-the-art report on alkali silica reactivity mitigation effectiveness using different types of fly ashes. *J. Mater.* **2016**, *2016*, 7871206. [CrossRef]
8. Thomas, M.D.A.; Fournier, B.; Folliard, K.J. *Selecting Measures to Prevent Deleterious Alkali-Silica Reaction in Concrete: Rationale for the AASHTO PP65 Prescriptive Approach*; Federal Highway Administration: Washington, DC, USA, 2012.
9. *ASTM C295*; Standard Guide for Petrographic Examination of Aggregates for Concrete. ASTM International: West Conshohocken, PA, USA, 2003.
10. *ASTM C1260*; Standard Test Method Potential Alkali React. Aggregates (Mortar-Bar Method). ASTM International: West Conshohocken, PA, USA, 2007.
11. *ASTM C 1293-08b*; Standard Test Method for Concrete Aggregates by Determination of Length Change of Concrete Due to Alkali-Silica Reaction. ASTM: West Conshohocken, PA, USA, 1995; p. 7.
12. Ichikawa, T. Alkali-silica reaction, pessimum effects and pozzolanic effect. *Cem. Concr. Res.* **2009**, *39*, 716–726. [CrossRef]
13. Diamond, S. A review of alkali-silica reaction and expansion mechanisms 1. Alkalies in cements and in concrete pore solutions. *Cem. Concr. Res.* **1975**, *5*, 329–345. [CrossRef]
14. Böhni, H. *Corrosion in Reinforced Concrete Structures*; Elsevier: Amsterdam, The Netherlands, 2005.
15. Diamond, S. A review of alkali-silica reaction and expansion mechanisms 2. Reactive aggregates. *Cem. Concr. Res.* **1976**, *6*, 549–560. [CrossRef]
16. Dent-Glasser, L.S. Osmotic Pressure and Swelling of Gels. *Cem. Concr. Res.* **1979**, *9*, 515–517. [CrossRef]
17. Dent-Glasser, L.S.; Kataoka, N. The Chemistry of AlkaliAggregate Reactions. In Proceedings of the Fifth International Conference on Alkali-Aggregate Reactions, Cape Town, South Africa, 30 March–3 April 1981; pp. 1–7.
18. Glasser, F.P. Chemistry of the Alkali-Aggregate Reaction. In *The Alkali-Silica Reaction in Concrete*; Swamy, R.N., Ed.; Van Nostrand Reinhold: New York, NY, USA, 1992; pp. 30–53. 333p.
19. Ichikawa, T.; Miura, M. Modified model of alkali-silica reaction. *Cem. Concr. Res.* **2007**, *37*, 1291–1297. [CrossRef]
20. Godart, B.; de Rooij, M.R.; Wood, J.G. *Guide to Diagnosis and Appraisal of AAR Damage to Concrete in Structures*; Springer: Berlin/Heidelberg, Germany, 2013.
21. Owsiak, Z. *Reakcje Kruszyw Krzemionkowych. Alkaliami w Betonie, Polski Biuletyn Ceramiczny Nr. 72*; Wydawnictwo Naukowe Akapit: Kraków, Poland, 2002. (In Polish)
22. Kurdowski, W. *Cement and Concrete Chemistry*; Springer: Dordrecht, The Netherlands, 2014.
23. Poyet, S.; Sellier, A.; Capra, B.; Foray, G.; Torrenti, J.M.; Cognon, H.; Bourdarot, E. Chemical modelling of alkali silica reaction: Influence of the reactive aggregate size distribution. *Mater. Struct.* **2007**, *40*, 229. [CrossRef]
24. Diamond, S. *Alkali Aggregate Reactions in Concrerete, an Annotated Bibliography 1939–1991*; National Academy of Sciences: Washington, DC, USA, 1992.

25. Swamy, R.N. Testing for Alkali Silica Reaction. In *The Alkali-Silica Reaction in Concrete*; Swamy, R.N., Ed.; Van Nostrand Reinhold: New York, NY, USA, 1992; pp. 54–95; 333p.
26. Moundoungou, I.; Bulteel, D.; Garcia-Diaz, E.; Thiéry, V.; Dégrugilliers, P.; Hammerschlag, J.G. Reduction of ASR expansion in concretes based on reactive chert aggregates: Effect of alkali neutralisation capacity. *Constr. Build. Mater.* **2014**, *54*, 147–162. [CrossRef]
27. Islam, M.S.; Akhtar, S. A critical assessment to the performance of alkali–silica reaction (ASR) in concrete. *Can. Chem. Trans.* **2013**, *1*, 253–266.
28. Broekmans, M.A.T.M. Structural properties of quartz and their potential role for ASR. *Mater. Charact.* **2004**, *53*, 129–140. [CrossRef]
29. Lindgård, J.; Andiç-Çakir, Ö.; Fernandes, I.; Rønning, T.F.; Thomas, M.D.A. Alkali-silica reactions (ASR): Literature review on parameters influencing laboratory performance testing. *Cem. Concr. Res.* **2012**, *42*, 223–243. [CrossRef]
30. Grattan-Bellew, P.E. Petrographic and Technological Methods for Evaluation of Concrete Aggregates. In *Handbook of Analytical Techniques in Concrete Science and Technology Principles, Techniques and Applications*; Ramachandran, V.S., Beaudoin, J.J., Eds.; Noyes Publication: Park Ridge, NJ, USA, 2001; pp. 63–98.
31. Bérubé, M.-A.; Fournier, B. Canadian experience with testing for alkaliaggregate reactivity in concrete. *Cem. Concr. Compos.* **1993**, *15*, 27–47. [CrossRef]
32. Farny, J.A.; Kosmatka, S.H. *Diagnosis and Control of Alkali-Aggregate Reactions in Concrete*; Concrete Information; Portland Cement Association: Washington, DC, USA, 1997; 23p.
33. ACI. *State of the Art Report on Alkali Aggregate Reactivity*; ACI Committee 221; ACI: Indianapolis, IN, USA, 1998; 31p.
34. Thomas, M.D.A.; Fournier, B.; Folliard, K.J.; Ideker, J.H.; Resendez, Y. *The Use of Lithium to Prevent or Mitigate Alkali-Silica Reaction in Concrete Pavementsand Structures*; Turner-Fairbank Highway Research Center: McLean, VA, USA, 2007.
35. Multon, S.; Cyr, M.; Sellier, A.; Diederich, P.; Petit, L. Effects of Aggregate Size and Alkali Content on ASR Expansion. *Cem. Concr. Res.* **2010**, *40*, 508–516. [CrossRef]
36. Stanton, T.E. *Expansion of Concrete through Reaction between Cement and Aggregate*; American Concrete Institute (ACI): Country Club Drive Farmington Hills, MI, USA, 2008; Available online: http://worldcat.org/isbn/9780870312694 (accessed on 7 February 2024).
37. Ben Haha, M.; Gallucci, E.; Guidoum, A.; Scrivener, K.L. Relation of expansion due to alkali silica reaction to the degree of reaction measured by SEM image analysis. *Cem. Concr. Res.* **2007**, *37*, 1206–1214. [CrossRef]
38. Kuroda, T.; Inoue, S.; Yoshino, A.; Nishibayashi, S. Effects of particle size grading and content of reactive aggregate on ASR expansion of mortars subjected to autoclave method. In Proceedings of the 12th International Conference on Alkali-Aggregate Reaction in Concrete, Beijing, China, 15–19 October 2004.
39. Wigum, B.J.; Lindgård, J. AAR: Testing, mitigation & recommendations. The Norwegian approach during two decades of research. In Proceedings of the 13th International Conference on Alkali-Aggregate Reaction in Concrete, Trondheim, Norway, 16–20 June 2008; pp. 1299–1309.
40. Saha, A.K.; Khan, M.N.N.; Sarker, P.K.; Shaikh, F.A.; Pramanik, A. The ASR mechanism of reactive aggregates in concrete and its mitigation by fly ash: A critical review. *Constr. Build. Mater.* **2018**, *171*, 743–758. [CrossRef]
41. Mehta, P.K.; Monteiro, P.J.M. *Concrete Microstructure, Properties and Materials*; Prentice-Hall Inc.: Englewood Cliffs, NJ, USA, 1993; 659p.
42. Helmuth, R.; Stark, D. Alkali-Silica Reactivity Mechanisms. In *Materials Science of Concrete III*; Skalny, F., Ed.; The American Ceramic Society: Westerville, OH, USA, 1992; pp. 131–138.
43. Zhang, C.; Wang, A.; Tang, M.; Wu, B.; Zhang, N. Influence of Aggregate Size and Aggregate Size Grading on ASR Expansion. *Cem. Concr. Res.* **1999**, *29*, 1393–1396. [CrossRef]
44. Ramyar, K.; Topal, A.; Andiç, Ö. Effects of Aggregate Size and Angularity on Alkali-Silica Reaction. *Cem. Concr. Res.* **2005**, *35*, 2165–2169. [CrossRef]
45. Andiç-Çakır, Ö. Investigation of Test Methods on Alkali Aggregate Reaction. Ph.D. Thesis, Ege University, Department of Civil Engineering, Bornova, Türkiye, 2007; 252p.
46. Chatterji, S. Chemistry of alkali-silica reaction and testing of aggregates. *Cem. Concr. Compos.* **2015**, *27*, 788–795. [CrossRef]
47. Islam, M.S.; Ghafoori, N. Relation of ASR-induced expansion and compressive strength of concrete. *Mater. Struct.* **2015**, *48*, 4055–4066. [CrossRef]
48. McNally, C.; Richardson, M.G. Reactivity assessment of aggregates: Role of chert crystallinity. *ACI Mater. J.* **2005**, *102*, 163–169.
49. Rajabipour, F.; Giannini, E.; Dunant, C.; Ideker, J.H.; Thomas, M.D.A. Alkali–silica reaction: Current understanding of the reaction mechanisms and the knowledge gaps. *Cem. Concr. Res.* **2015**, *76*, 130–146. [CrossRef]
50. Aşık, İ.; Şen, H.; Ergintav, Y.; Ünsal, A.; Şentürk, E.; Bayrak, E. *Rehabilitation of the Agregate Harmful in Respect of Alkali Silica Reation*; Concrete Congress: Ankara, Turkey, 2004.
51. Alexander, M.G.; Mindness, S. *Aggregates in Concrete (Modern Concrete Technology)*; CRC Press: Boca Raton, FL, USA, 2005; 448p.
52. Thomas, M.D.A.; Fournier, B.; Folliard, K.J.; Resendez, Y. *Alkali-Silica Reactivity Field Identification Handbook*; Federal Highway Administration, Office of Pavement Technology: Washington, DC, USA, 2011.
53. Lima, M.; Salamy, R.; Miller, D. CFRP strengthening of ASR affected concrete piers of railway bridges. In *Maintenance, Safety, Risk, Manag. Life-Cycle Performance of Bridges, Proceedings of the Ninth International Conference on Bridge Maintenance, Melbourne, Australia, 9–13 July 2018*; CRC Press: Boca Raton, FL, USA, 2018; p. 346.

54. Fanijo, E.O.; Kolawole, J.T.; Almakrab, A. Alkali-silica reaction (ASR) in concrete structures: Mechanisms, effects and evaluation test methods adopted in the United States. *Case Stud. Constr. Mater.* **2021**, *15*, e00563. [CrossRef]
55. Deschenes, D.J.; Bayrak, O.; Folliard, K.J. *ASR/DEF-Damaged Bent Caps: Shear Tests and Field Implications*; Technical Report No. 12-18XXIA006; University of Texas: Austin, TX, USA, 2009; 258p.
56. Snyder, K.A.; Lew, H.S. *Alkali-Silica Reaction Degradation of Nuclear Power Plant Concrete Structures: A Scoping Study*; National Institute of Standards and Technology Engineering Laboratory: Gaithersburg, MD, USA, 2013.
57. Bektasa, F.; Turanlia, L.; Topal, T.; Goncuoglu, M.C. Alkali reactivity of mortars containing chert and incorporating moderate-calcium fly ash. *Cem. Concr. Res.* **2004**, *34*, 2209–2214. [CrossRef]
58. Williams, H.; Turner, F.J.; Gilbert, C.H. *Petrography*; Freeman: New York, NY, USA, 1982.
59. Wigum, B.J. Alkali–Aggregate Reactions in Concrete: Properties, Testing and Classification of Norwegian Cataclastic Rocks. Ph.D. Thesis, University of Trondheim, Trondheim, Norway, 1995.
60. Richardson, M. Minimising the risk of deleterious alkali–silica reaction in Irish concrete practice. *Constr. Build Mater.* **2005**, *19*, 654–660. [CrossRef]
61. Nishiyama, T.; Kusuda, H.; Nakano, K.A. Few remarks on alkalireactive chert aggregates. In Proceedings of the Eighth International Conference on Alkali-Aggregate Reaction, Kyoto, Japan, 17–20 July 1989; Okada, K., Nishibayashi, S., Kawamura, M., Eds.; pp. 543–548.
62. Jones, T.N. Mechanism of reactions involving British chert and flint aggregates. In Proceedings of the Eighth International Conference on Alkali-Aggregate Reaction, Kyoto, Japan, 17–20 July 1989; Okada, K., Nishibayashi, S., Kawamura, M., Eds.; pp. 135–140.
63. Larbi, J.A.; Visser, J.H.M. A study of the ASR of an aggregate with high chert content by means of ultra accelerated mortar bar test and pore fluid analysis. *Heron* **2002**, *47*, 141–159.
64. Gogte, B.S. An evaluation of some common Indian rocks with special reference to alkali aggregate reactions. *Eng. Geol.* **1973**, *7*, 135–154. [CrossRef]
65. Fournier, B.; Chevier, R.; De Grosbois, M.; Lisella, R.; Folliard, K.; Idekar, J.; Shehata, M.; Thomas, M.; Baxer, S. The Accelerated Concrete Prism Test (60 °C): Variability of the Test Method and Proposed Expansion Limits. In Proceedings of the 12th International Conference on Alkali-Aggregate Reaction in Concrete, Beijing, China, 15–19 October 2004; Tang, M., Deng, M., Eds.; World Publishing Corporation: Beijing, China, 2004; pp. 314–323.
66. *ASTM C 597*; Designation: C597–09 Standard Standard Test Method for Pulse Velocity Through Concrete. ASTM International: West Conshohocken, PA, USA, 2010.
67. *TS EN 196-1*; Çimento Deney Yöntemleri, Dayanım Tayini. Türk Standartları Enstitüsü: Ankara, Turkey, 2002.
68. *TS EN 196-2*; Çimento Deney Yöntemleri—Bölüm 2: Çimentonun Kimyasal Analizi. Türk Standartları Enstitüsü: Ankara, Turkey, 2013.
69. *ASTM C 150*; Standard Specification for Portland Cement. ASTM: West Conshohocken, PA, USA, 2005; p. 8.
70. Wang, H.; Gillott, J.E. Effect of some chemicals on alkali silicareaction. In Proceedings of the 9th International Conference on Alkali-Aggregate Reaction, London, UK, 27–31 July 1992; Concrete Society of UK: London, UK, 1992; pp. 1090–1099.
71. Yalçın, H.; Özalp, R. *Alkali-Agrega Reaction in Concrete and Alkalinity Values of Turkish Cements*; State Hydraulic Works Publication: Ankara, Turkey, 1974.
72. Mindess, S.; Young, J.F. *Concrete*; Prentice-Hall, Inc.: Hoboken, NJ, USA, 1981.
73. Xu, H. On the alkali content of cement in AAR, Concr. Alkali-Aggregate React. In *Proceedings of the 7th International Conference*; Grattan-Bellew, P.E., Ed.; Noyes Publications: Park Ridge, NJ, USA, 1987; pp. 451–455.
74. Bérubé, M.-A.; Frenette, J. Testing concrete for AAR in NaOH and NaCl solutions at 38 C and 80 C. *Cem. Concr. Compos.* **1994**, *16*, 189–198. [CrossRef]
75. Ramyar, K.; Dönmez, H.; Andiç, Ö. *Control of Alkali Silica Reaction by Mineral and Chemical Additives*; Turkish Cement Manufacturers' Association, Cement and Concrete Research and Development Institute: Ankara, Turkey, 2002.
76. Da Thomas, M.; Fournier, B.; Folliard, K.J. *Alkali-Aggregate Reactivity (AAR) Facts Book*; Federal Highway Administration, Office of Pavement Technology: Washington, DC, USA, 2013.
77. Woods, H. *Durability of Concrete Construction, Monograph No. 4*; American Concrete Institute: Detroit, MI, USA, 1968.
78. SHRP. Alkali Silica Reactivity Library Hanbook for Identification ASR. U.S. Department of Transportation Federal Aviation Administration, 2004. Available online: https://leadstates.tamu.edu/asr (accessed on 7 February 2024).
79. Nishibayashi, S.; Yamura, K. Effect of reactive fine aggregate on expansion characteristics of concrete due to alkali aggregate reaction. In Proceedings of the 9th International Conference on Alkali-Aggregate Reaction in Concrete, London, UK, 27–31 July 1992; pp. 723–730.
80. Sims, I.; Nixon, P. RILEM recommended test method AAR-1: Detection of potential alkali reactivity of Aggregates-Petrographic method. *Mater. Struct.* **2003**, *36*, 480–496. [CrossRef]
81. Binal, A. The investigation of effects of pessimum reactive aggregate content on alkali-silica reaction with experimental methods. *Istanb. Univ.-Turk. Eng. Fak. J. Earth Sci.* **2004**, *17*, 119–128.
82. Eker, H.; Demir Şahin, D.; Çullu, M. Effect of Reduced Fineness of Fly Ash Used on the Alkali–Silica Reaction (ASR) of Concrete, Iranian Journal of Science and Technology. *Trans. Civ. Eng.* **2023**, *47*, 2203–2217. [CrossRef]
83. Whitehurst, E.A. Soniscope Test Concrete Structures. *J. Am. Concr. Inst. Proc.* **1951**, *47*, 443–444.

84. Uyanık, O.; Kaptan, K.; Gülay, F.G.; Tezcan, S. Determination of Concrete Strength by Non-Destructive Ultrasonic Method. *World Constr.* **2011**, *184*, 55–58.
85. Demir, İ. The mechanical properties of alkali-silica reactive mortars containing same amounts of silica fume and fly ash. *Gazi Univ. J. Fac. Eng. Archit.* **2010**, *25*, 749–758.
86. Swamy, R.N.; Al-Asali, M.M. Expansion of concrete due to ASR. *ACI Mater. J.* **1988**, *85*, 33–40.
87. Neville, A.M. *Properties of Concrete*; Longman Scientific & Technical: Harlow, UK, 1981.
88. Gümüş, S. Determining the Alkali—Agrega Reactivitiy of the Aggregates at Kocaelđ Region. Master's Thesis, Sakarya University Institute of Science and Technology, Sakarya, Türkiye, 2009.

buildings

Article

Stabilization Effects of Inclined Soil–Cement Continuous Mixing Walls for Existing Warm Frozen Soil Embankments

Gaochen Sun [1,*], Long Li [1], Yufan Huo [1], Zonghui Fang [2], Gao Lv [1], Tao Chen [1], Zhengzhong Wang [3], Meimei Song [1], Guanqi Lan [1] and Ziqiang Tang [4]

1 School of Civil Engineering, Xi'an Shiyou University, Xi'an 710065, China; 17730623101@163.com (L.L.); 13379589757@163.com (Y.H.); lvgao116@163.com (G.L.); chentao@163.com (T.C.); songmeimei2016@163.com (M.S.); langq@xsyu.edu.cn (G.L.)
2 Gansu Institute of Architecture Design and Research Co., Ltd., Lanzhou 730030, China; 18298385727@163.com
3 College of Water Resources and Architectural Engineering, Northwest A&F University, Yanglin 712100, China; wangzz0910@163.com
4 China State Construction Engineering Corporation Aecom Consultant Co., Ltd., Lanzhou 730030, China; show-19870629@163.com
* Correspondence: maomaosgc@163.com

Abstract: Affected by climate warming and anthropogenic disturbances, the thermo-mechanical stability of warm and ice-rich frozen ground along the Qinghai–Tibet Railway (QTR) is continuously decreasing, and melting subsidence damage to existing warm frozen soil (WFS) embankments is constantly occurring, thus seriously affecting the stability and safety of the existing WFS embankments. In this study, in order to solve the problems associated with the melting settlement of existing WFS embankments, a novel reinforcement technology for ground improvement, called an inclined soil–cement continuous mixing wall (ISCW), is proposed to reinforce embankments in warm and ice-rich permafrost regions. A numerical simulation of a finite element model was conducted to study the freeze–thaw process and evaluate the stabilization effects of the ISCW on an existing WFS embankment of the QTR. The numerical investigations revealed that the ISCW can efficiently reduce the melt settlement in the existing WFS embankment, as well as increase the bearing capacity of the existing WFS embankment, making it favorable for improving the bearing ability of composite foundations. The present investigation breaks through the traditional ideas of "active cooling" and "passive protection" and provides valuable guidelines for the choice of engineering supporting techniques to stabilize existing WFS embankments along the QTR.

Keywords: existing warm frozen soil (WFS) embankment; melt subsidence disaster; inclined soil–cement mixing continuous wall (ISCW); numerical simulation; melt settlement

Citation: Sun, G.; Li, L.; Huo, Y.; Fang, Z.; Lv, G.; Chen, T.; Wang, Z.; Song, M.; Lan, G.; Tang, Z. Stabilization Effects of Inclined Soil–Cement Continuous Mixing Walls for Existing Warm Frozen Soil Embankments. *Buildings* **2024**, *14*, 1892. https://doi.org/10.3390/buildings14071892

Academic Editors: Huayang Yu and Fabrizio Gara

Received: 18 April 2024
Revised: 5 June 2024
Accepted: 12 June 2024
Published: 21 June 2024

1. Introduction

China is the third largest country with frozen soil in the world, with the frozen soil area accounting for 68.6% of its total land area [1,2]. With the warming of the global climate, permafrost is gradually transforming into warm frozen soil, and the warm frozen soil (WFS) embankment phenomenon continues to emerge. WFS, also known as plastic frozen soil, is a special frozen soil body undergoing a drastic phase change with a temperature of about −1 °C. It is characterized by high water content, high compressibility, low strength, and poor engineering properties [3,4]. Studies have shown that, in sections with large amounts of WFS, the proportion of embankment diseases is increased. According to experience from the Qinghai–Tibet Railway (QTR), melt subsidence of an embankment is the main disease in permafrost areas [5,6]. In particular, when an embankment begins to melt as the temperature rises in the spring, the soil layer under the pavement melts faster than the soil layer under the road shoulder, forming a concave impervious frozen soil core in the middle of the embankment, which cannot discharge the water of the melted upper soil mass, causing the embankment to be damaged by churning [7]. It has been found that longitudinal cracks on high embankments in WFS areas are caused by different melting and settling of shaded and sunny slopes. In addition, embankments in WFS areas are prone to settlement, mainly because of the compression deformation and melting consolidation in the WFS layer [8,9]. The continuous generation of WFS has aggravated the occurrence of melt subsidence disasters on the Qinghai–Tibet Railway embankments in frozen soil areas, threatening their safe operation.

In order to reduce the occurrence of melt subsidence disasters, traditional permafrost subgrade insulation measures follow the principles of "active cooling" and "passive protection" [10,11]; however, these measures can only be applied in an environment where future temperature increases remain within 1 °C. With global warming, the protective effect on an existing embankment will be weakened, the effect on the WFS area will not be obvious, and the WFS area will continue to increase. Therefore, these methods also have certain limitations, and there is an urgent need to develop new methods for WFS embankment reinforcement.

Soil–cement mixing walls are often used to strengthen soft soil foundations. The key principle is to use cement as a curing agent and mix the cement with soft soil deep in an embankment using mixing machinery to form a soil–cement mixing wall [12]. The cement hydration reaction with free water in the soft soil changes the physical and mechanical properties of the soft soil and increases its bearing capacity [13,14]. A soil–cement mixing wall used as a composite foundation bears most of the load, which can improve the working performance of the soft soil foundation more than other stabilized methods [15,16]. Through a literature review, we found that the physical and mechanical properties of WFS are similar to soft soil, in terms of their characteristics of high moisture content, low shear resistance, and high compressibility [17]. Some previous studies have shown that adding cement to WFS can alter its characteristics, thus reinforcing the strength of the WFS [18–20]. In permafrost regions that face existing embankment thaw subsidence problems, it is difficult to take advantage of the other techniques for existing embankment stabilization. Hence, through combining the soil–cement mixing wall method with the inclined piles concept, we derive a novel WFS embankment improvement technique, called inclined soil–cement continuous mixing wall (ISCW) installation, for engineering construction in warm and ice-rich permafrost regions.

In this study, the inclined soil–cement continuous mixing wall (ISCW) method was considered for WFS embankments of the QTR presenting the thaw subsidence process, and the associated stabilized effects were numerically simulated. The effects of the ISCW, in terms of stabilizing the WFS embankment, were analyzed by combining thermal and deformation processes. The specific objective of this study is to provide a novel method to reduce the threat of melt subsidence disasters in existing WFS embankments, thus supporting the infrastructural integrity of major projects along the Qinghai–Tibet engineering corridor (QTEC) in the future.

2. Numerical Model of the ISCW for Existing WFS Subgrade

2.1. Assumption

Considering the complexity of the foundation environment of warm frozen soil, a simplified consolidation model is adopted based on the following assumptions:

(1) The soil in each layer is continuous, homogeneous, and isotropic, and the soil is an ideal elastic–plastic material.
(2) In the calculation, only the deformation of the embankment caused by the applied load is considered, and the action of gravity is not considered.
(3) The stress and displacement of soil in the process of pile formation are not considered.
(4) The deformation and contact between the soil layers are coordinated

2.2. Heat Equations for Frozen Soil

The calculation model of heat conduction in frozen soil can be divided into freezing and melting states (FMSs). Hence, the freezing and melting states equation for the heat conduction of frozen soil can be expressed as Equations (1) and (2) [21]:

(1) Freeze state ($T_s = T_f$):

$$C_f\frac{\partial T_f}{\partial t} = \frac{\partial}{\partial x}(\lambda_f\frac{\partial T_f}{\partial x}) + \frac{\partial}{\partial y}(\lambda_f\frac{\partial T_f}{\partial y}) + \frac{\partial}{\partial z}(\lambda_f\frac{\partial T_f}{\partial z}) \quad (1)$$

(2) Melt state ($T_s = T_u$):

$$C_u\frac{\partial T_u}{\partial t} = \frac{\partial}{\partial x}(\lambda_u\frac{\partial T_u}{\partial x}) + \frac{\partial}{\partial y}(\lambda_u\frac{\partial T_u}{\partial y}) + \frac{\partial}{\partial z}(\lambda_u\frac{\partial T_u}{\partial z}) \quad (2)$$

where T_s represents the FMS temperature; T_f represents the freezing temperature; T_u represents the melting temperature; C_f represents the volumetric heat capacity in the freezing state; λ_f represents the thermal conductivity in the freezing state; C_u represents the volumetric heat capacity in the melting state; and λ_u represents the volumetric heat in the melting state.

The phase change is assumed to occur in a temperature range of $T_m \pm \Delta T$. Therefore, the heat capacity C_s and the thermal conductivity λ_s of the frozen soil are given by Equations (3) and (4), respectively:

$$C = \begin{cases} C_f & T < T_m - \Delta T \\ C_f + \frac{C_t - C_f}{T_a - T_b}(T - \Delta T) + \frac{L}{1+W}\frac{\partial W_i}{\partial T} & T_m - \Delta T \le T \le T_m - \Delta T \\ C_t & T > T_m + \Delta T \end{cases} \quad (3)$$

$$\lambda = \begin{cases} \lambda_f & T < T_m - \Delta T \\ \lambda_f + \frac{\lambda_t - \lambda_f}{T_a - T_b}(T - T_m + \Delta T) & T_m - \Delta T \le T \le T_m + \Delta T \\ \lambda_u & T > T_m + \Delta T \end{cases} \quad (4)$$

where T_m is the freezing temperature and ΔT is the temperature increment.

Therefore, Equations (3) and (4) can be simplified to (Equation (5)):

$$C_s\frac{\partial T_s}{\partial t} = \frac{\partial}{\partial x}(\lambda_s\frac{\partial T_s}{\partial x}) + \frac{\partial}{\partial y}(\lambda_s\frac{\partial T_s}{\partial y}) + \frac{\partial}{\partial z}(\lambda_s\frac{\partial T_s}{\partial z}) \quad (5)$$

2.3. Thermodynamic Coupling Governing Equation

The stress field under thermodynamic coupling is a plane strain problem, and its governing equations [22] are shown in Equations (6) and (8) as follows:

$$\varepsilon_x = \frac{1-\mu^2}{E(T)}(\sigma_x - \frac{\mu}{1-\mu}\sigma_y) + \alpha\Delta T \quad (6)$$

$$\varepsilon_y = \frac{1-\mu^2}{E(T)}(\sigma_y - \frac{\mu}{1-\mu}\sigma_x) + \alpha\Delta T \quad (7)$$

$$\varepsilon_{xy} = \frac{2(1-\mu)}{E(T)}\tau_{xy} \quad (8)$$

where σ_x and σ_y are the *x*- and *y*-direction normal stresses; ΔT is the temperature gradient; ε_x and ε_y are the *x*- and *y*-direction plane shear strains; ε_{xy} is the plane shear strain in the *xy*-plane; μ is Poisson's ratio; α is the coefficient of thermal expansion; and $E(T)$ is the elastic modulus of a material.

2.4. Constitutive Model

In this study, both the ISCW and the WFS included in the embankment and foundation were considered as elastoplastic materials. The stress–strain relationship is given by the following increment formulation (Equation (9)) [23]:

$$\{d\sigma\} = \{D_T\}\left[\{d\varepsilon\} - \{d\varepsilon_p\}\right] \tag{9}$$

where $\{d\sigma\}$ is the incremental stress matrix; $\{D_T\}$ is the elastic matrix depending on temperature, which can be expressed by E_T (elastic modulus) and v_T (Poisson's ratio); $\{d\varepsilon\}$ is the total incremental strain matrix; and $\{d\varepsilon_p\}$ is the plastic incremental strain matrix.

According to the plastic potential theory, the plastic strain increment of materials can be formed. For elastoplastic materials, this is related to the stress state and plastic strain of the plastic potential function H. Therefore, the plastic strain increment is given as follows (Equation (10)) [24]:

$$\{d\varepsilon_p\} = d\lambda\frac{\partial H}{\partial\{\sigma\}} \tag{10}$$

Equation (10) determines the relationship between the plastic strain increment and stress increment within the plastic range of geotechnical materials, where $d\lambda$ represents the plastic multiplier.

The elastoplastic model with the Drucker–Prager yield criterion in coordination with the Mohr–Coulomb yield criterion was used in this study. Applying the relevant flow rules (the plastic potential function H is the same as the subsequent yield surface function F), Equation (10) can be rewritten as Equations (11)–(13):

$$\{d\varepsilon_p\} = d\lambda\frac{\partial F}{\partial\{\sigma\}} \tag{11}$$

$$\frac{\partial F}{\partial\{\sigma\}} = \begin{bmatrix} \frac{\partial F}{\partial\sigma_r} & \frac{\partial F}{\partial\sigma_z} & \frac{\partial F}{\partial\sigma_{rz}} & \frac{\partial F}{\partial\sigma_\theta} \end{bmatrix}^T \tag{12}$$

$$F = \sqrt{J_2} + \alpha I_1 - k \tag{13}$$

where J_2 is the second-order deviatoric stress invariant, I_1 is the first stress tensor invariant, and k are material parameters related to the cohesion c and the internal friction angle φ variables, defined according to Equations (14) and (15), respectively:

$$\alpha = \frac{2\sin\varphi}{\sqrt{3}(3-\sin\varphi)} \tag{14}$$

$$k = \frac{2\sqrt{3}c\cos\varphi}{3-\sin\varphi} \tag{15}$$

3. Numerical Simulations

3.1. Model Prototype

According to Reference [25] and ISCW design specifications, an unstable embankment in the QTR was selected to establish the calculation model. The embankment top has a width of 8.0 m and a height of 5.0 m. The calculated boundary width of the embankment foundation is 23 m, and the calculated boundary distance between the toes of the embankment is 18.5 m. The foundation depth is 20 m. A cross-section of the model prototype is shown in Figure 1.

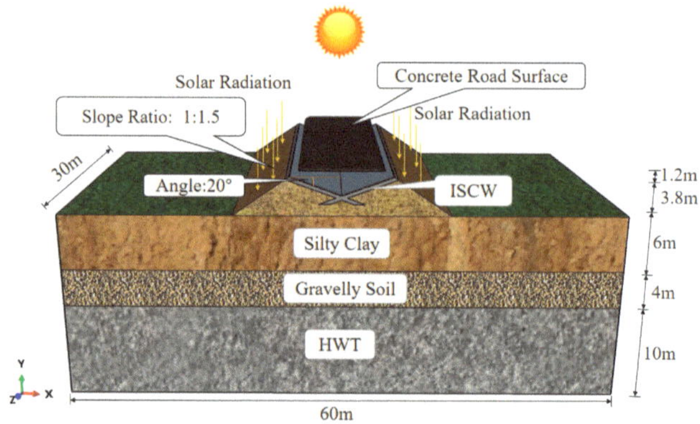

Figure 1. ISCW composite embankment of the WFS embankment model prototype.

A thermo-mechanical coupling finite element analysis model was established, and the size effect was considered. The length, width, and height of the model were 60 m, 30 m, and 20 m, respectively. The embankment height was 5.0 m, the road width was 8 m, and the slope was 1:1.5. The length of the ISCW was l = 7 m, the thickness was 0.4 m, and the angle between the diaphragm wall and the horizontal plane was 20°. The top of the diaphragm wall was 1.2 m from the road. C3D8R hexahedral elements were used in the grid, and the total number of elements was 43,320, which was calculated using reduced integrals, as shown in Figure 1. The embankment was divided into concrete pavement (0.3 m), embankment fill (4.7 m), silty clay (6.0 m), gravel soil (4.0 m), and strongly weathered tuff (10.0 m) from top to bottom. The model cross-section is shown in Figure 1, and the thermodynamic parameters of the model are provided in Table 1.

Table 1. Physical parameters of the constructed model.

Material	ρ (kg/m³)	λ_u/(w/m·°C)	C_u/(J/kg·°C)	*E*/pa	μ	*φ*/(°)	*C*/(pa)	*H*/m
Concrete road surface	2400	1.6	976	30×10^9	0.2	/	/	0.3
Embankment	1900	1.82	1463	20×10^6	0.3	16	12,000	4.7
Silty clay	1800	1.24	2090	65×10^6	0.3	11	55,000	6
Gravelly soil	1900	1.63	1560	85×10^6	0.3	25	50,000	4
HWT	2000	1.6	877	110×10^6	0.25	20	55,000	10
ISCW	2200	1.2	1580	5×10^9	0.25	/	/	7

ρ, dry density; λ_u, heat conductivity; c_u, heat coefficient; *E*, elastic modulus; μ, Poisson's ratio; *C*, friction angle; *φ*, cohesion; *H*, soil thickness; HWT, highly weathered tuff; ISCW, inclined soil–cement continuous wall.

The calculation of the thermodynamic parameters for each material was based on previous research [26] and the literature such as the Frost Action of Soil and Foundation Engineering [27], Frozen Soil Physics [28], Code for Design of ISCW (Standards Press of China 2010) [29], and Code for Thermal Design of ISCW (China Academy of Building Research 1993) [30]. The values for the materials' mechanical parameters were adopted from the literature [31].

3.2. Finite Element Model

ABAQUS (2022) is a software program used for geotechnical material and other material performance analysis [32]. Because of its command-driven mode, specificity, and openness, ABAQUS has been widely used in the study of soil failure, collapse, and soil consolidation [33]. In this study, the ABAQUS finite difference method was used to construct the ISCW on the existing WFS subgrade and observe the stabilization effect of the ISCW on the subgrade.

The existing embankment, ISCW, and all soil layers were fine-meshed using hexahedral elements, and nodal points were connected among adjacent solid elements automatically. Figure 2 illustrates the work tree of mesh generation for the embankment, ISCW, and foundation in this study.

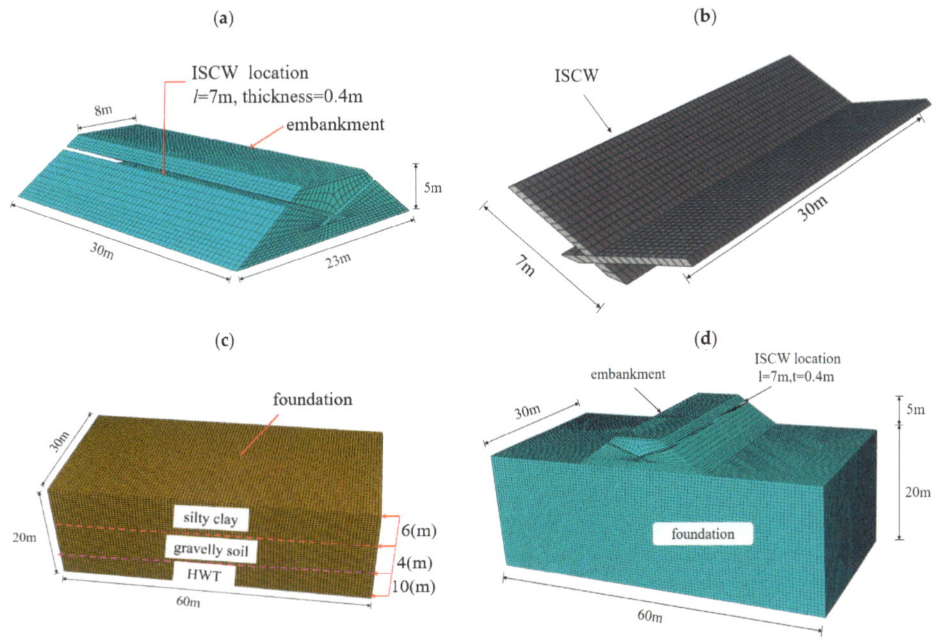

Figure 2. ISCW composite embankment of the WFS embankment finite element model. (**a**) Embankment; (**b**) ISCW; (**c**) foundation; and (**d**) finite element model.

3.3. Boundary and Initial Conditions

3.3.1. Temperature Field Analysis Steps and Boundary Conditions

In the heat transfer analysis step, the upper surface boundary condition was set as the local surface temperature in December, the shade slope surface was set at −15.7 °C, the central surface of the line was set at −12.0 °C, the sunny slope surface was set at −11.0 °C, and the natural surface temperature was set at −14.1 °C. The lower surface boundary condition was set as a constant temperature condition of −2 °C [34]. According to the measured data and considering the influence of geothermal energy, the initial temperature was set as the heat insulation condition around the surface, which was used as the initial condition for the subsequent temperature field calculation.

The initial conditions of the stress field were determined in the heat transfer analysis step, and the time length was 120. The K3016 section near Wudaoliang in the QTR was selected as the research object, and the surface temperature during the four months from December to March, for a total of 120 days, was taken as the boundary condition of the temperature field. The parameters of the temperature field for the boundary conditions are detailed in Table 2 [35]:

Considering solar radiation, previous studies [36] used a periodic function to fit the land surface temperature and adopted the simplified boundary temperature condition of the cross-section shown in Figure 1. The fitting formula is shown in Equation (16):

$$T_s = T_0 + A_0 \sin(\omega t + \varphi) \tag{16}$$

Table 2. Parameter assignment for the surface sine temperature function.

T_0/(°C)	A_0/(°C)	φ/(rad)	Surface
−1.4	12.8	11π/18	Original surface
−0.3	11.3	5π/9	South slope
−3.0	13.1	21π/36	North slope
2.6	14.8	11π/18	Ground

T_0, annual mean temperature; A_0, amplitude; φ, phase angle.

3.3.2. Stress Field Analysis Steps and Boundary Conditions

The static general analysis step was adopted, and the time length was 120. The upper boundary of the model was not constrained; the left and right boundary constraint method was phase shift; and the lower boundary constraints were horizontal displacement and vertical displacement.

3.3.3. Contact Element Setting

The surface of the pile sides and bottom was set as the secondary surface, and the soil surface was set as the primary surface. The contact property between the pile and soil was set as follows: the friction formula of tangential behavior was set as the penalty function, where the friction coefficient was 0.5; normal behavior was set to "hard" contact.

4. Results and Analysis

4.1. Temperature Analysis for the Reinforced and Unreinforced Embankments

As the temperature increased, the temperature of the road surface and slope also increased. Because of the effects of shade and sun on the slope, the right-side slope reached the sunny temperature first, and the temperature continued to transfer to the un-melted soil in the lower part of the embankment. As can be seen in Figure 3, melting mainly occurred in the shallow soil of the embankment, and the temperature of the deep soil remained basically stable at about −2 °C. Four months after melting, the lowest temperature of the unreinforced embankment was −11.1 °C, and the lowest temperature of the composite embankment was −13.5 °C; therefore, the lowest temperature of the composite embankment was 2.4 °C lower than that of the unreinforced embankment. The temperature calculation results for the bottom surface of the embankment four months after melting are shown in Figure 4a. The lowest temperature of the composite embankment was 2.1 °C lower than that of the unreinforced embankment. As can be seen in Figure 4b, in the range of 0 m to −5 m, the temperature of the composite embankment was lower than that of the unreinforced embankment, and the peak value of the negative temperature occurred. In the range of −5 m to −20 m, the temperature of the composite embankment was higher than that of the unreinforced embankment. This is because the continuous wall under the cement can absorb the negative temperature of the outside world in the cold season and transfer its negative temperature to the embankment, causing the temperature of the embankment to decrease. With an increase in external temperature, the embankment gradually melts, and the continuous wall under the cement land can isolate the temperature of the upper embankment, transferring the negative temperature of the deep soil under the embankment to the upper embankment, thereby reducing the temperature of the embankment. As the temperature of the cement–soil continuous wall was lower than that in other parts, a peak was observed on the temperature curve. The ISCW has the function of reducing the temperature and slowing down the melting settlement of the embankment.

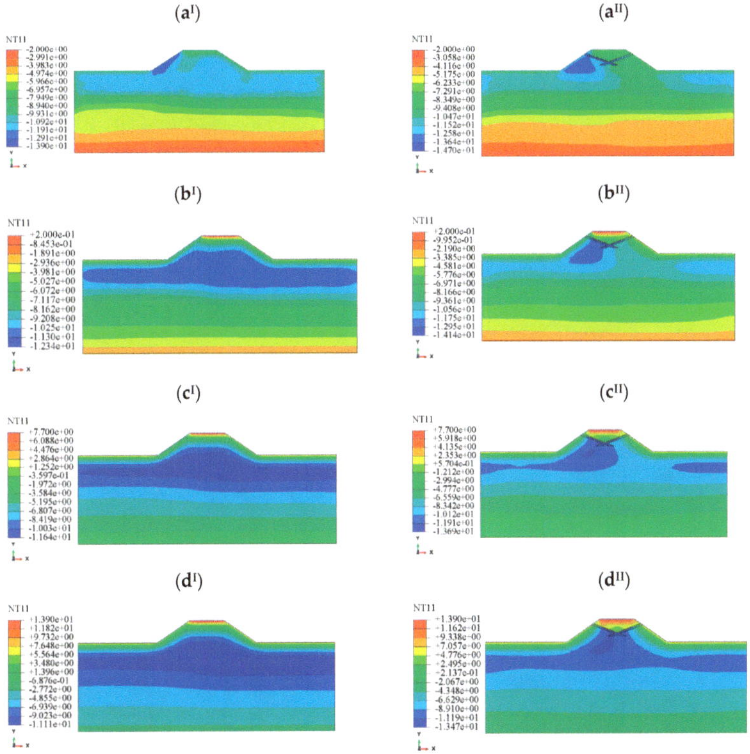

Figure 3. Temperature field cloud image before and after reinforcement (unit: °C). Unreinforced embankment. (**a^{I}**) First month; (**b^{I}**) second month; (**c^{I}**) third month; and (**d^{I}**) fourth month. Reinforced embankment. (**a^{II}**) First month; (**b^{II}**) second month; (**c^{II}**) third month; and (**d^{II}**) fourth month.

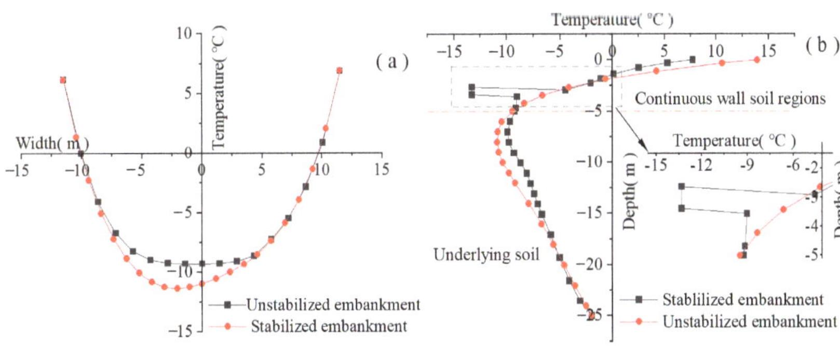

Figure 4. Results of the thermal calculation for the reinforced and unreinforced embankments. (**a**) Bottom of the embankment and (**b**) the center line temperature of the embankment along the *y*-direction after melt.

4.2. Vertical Displacement (Y-Displacement) Analysis of the Reinforced and Unreinforced Embankments

The vertical displacement in the reinforced and unreinforced embankments is shown in Figure 5, from which it can be clearly seen that the settlement range of the composite embankment was larger than that of the unreinforced embankment in the fourth month. This is because the composite embankment is formed from the underground continuous wall and the soil mass, and the load on the road is transferred to the underground continuous wall to jointly bear the load, thus reducing the displacement in the vertical direction. Figure 6 shows the calculation results of the displacement in the vertical direction for the embankment top surface during the four-month period over which the embankment melted. The maximum displacement in the vertical direction for the unreinforced embankment increased from 0.34 cm to 2.37 cm, and the displacement increased by 1.96 cm. The maximum displacement in the y-direction for the composite embankment increased from 0.33 cm to 1.71 cm, and the displacement increased by 1.38 cm. The maximum displacement of the composite embankment was 28% lower than that of the unreinforced embankment. Therefore, the use of a continuous wall under inclined cement soil can effectively reduce the melt displacement of the embankment.

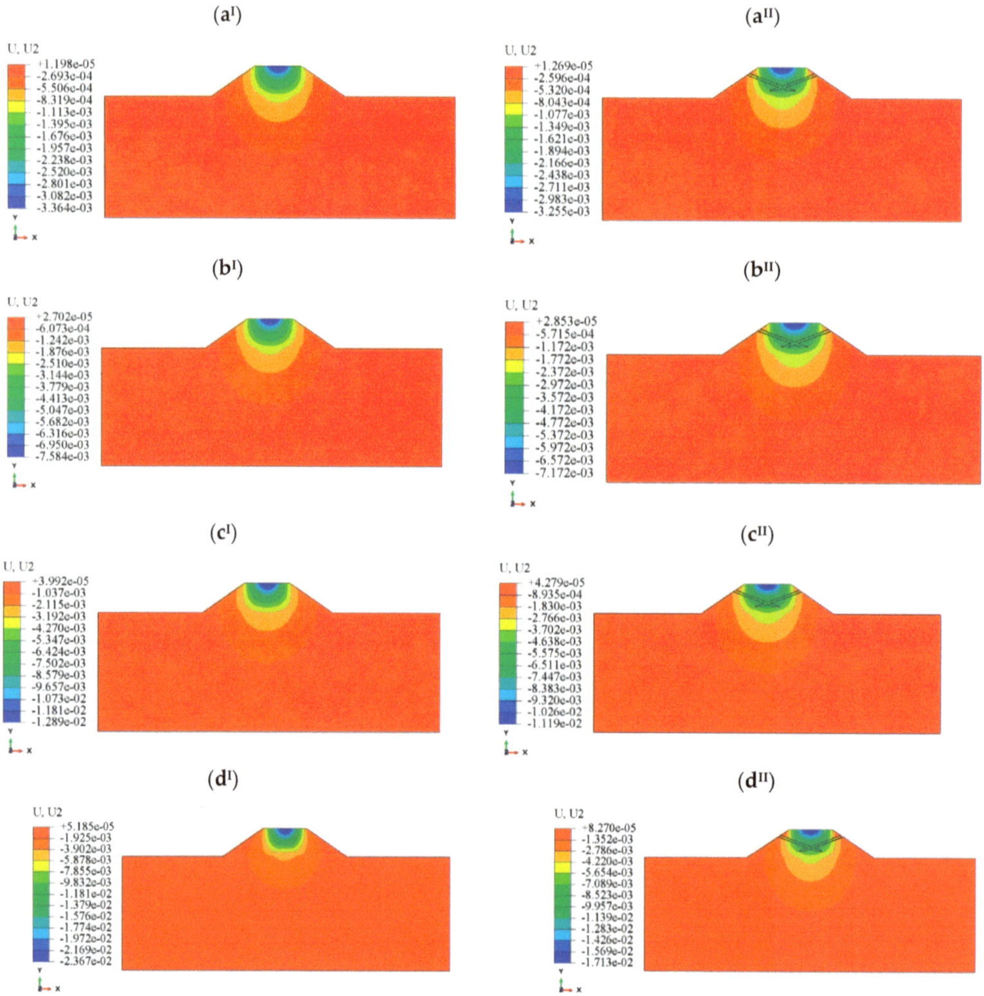

Figure 5. Cloud image of reinforcement before and after vertical displacement (Unit: m). Unreinforced embankment. (**a^{I}**) First month; (**b^{I}**) second month; (**c^{I}**) third month; and (**d^{I}**) fourth month. Reinforced embankment. (**a^{II}**) First month; (**b^{II}**) second month; (**c^{II}**) third month; and (**d^{II}**) fourth month.

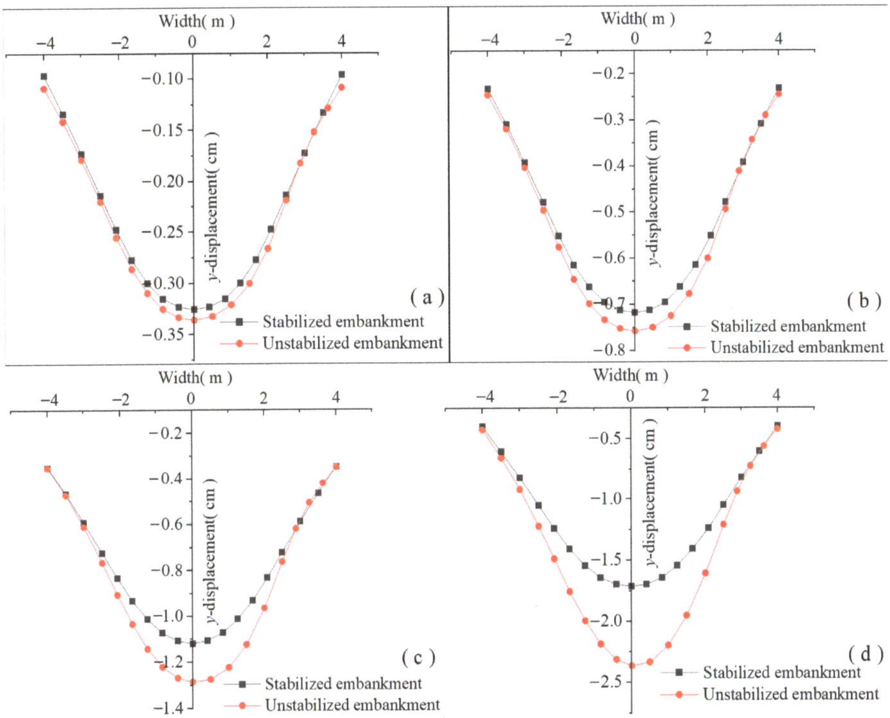

Figure 6. Monthly vertical displacement (Y-displacement) comparison in embankment surfaces before and after consolidation. (**a**) First month; (**b**) second month; (**c**) third month; and (**d**) fourth month.

4.3. Horizontal Displacement (X-Displacement) Analysis of the Reinforced and Unreinforced Embankments

In Figure 7, it can be seen that the maximum displacement in the horizontal direction for the unreinforced embankment ranged from 0.43 mm to 5.1 mm, and the displacement increased by 4.67 mm. The maximum displacement in the horizontal direction of the composite embankment increased from 0.59 mm to 3.57 mm, and the displacement increased by 2.98 mm. Therefore, the maximum displacement of the composite embankment was 30% less than that of the unreinforced embankment. Figure 8 shows the calculation results for the displacement of the embankment in the horizontal direction at a distance of 3.8 m from the top of the embankment during the four-month period over which the embankment melted. During the four-month period, the displacement of the unreinforced embankment in the x-direction increased from 0.33 mm to 3.44 mm, and the displacement increased by 3.11 mm. The displacement in the x-direction of the composite embankment increased from 0.25 mm to 1.0 mm, and the displacement increased by 0.75 mm. Therefore, the displacement of the composite embankment was 67.8% lower than that of the unreinforced embankment. The deformation of the embankment is mainly due to the load on the road surface. The ISCW can interact with the soil to form a composite embankment, which increases the integrity of the embankment. Moreover, the ISSCW has a "clamping" effect on the soil, and the maximum horizontal displacement occurs in the continuous wall, which greatly reduces the horizontal displacement of the embankment below the continuous wall, thus reducing the deformation of the embankment.

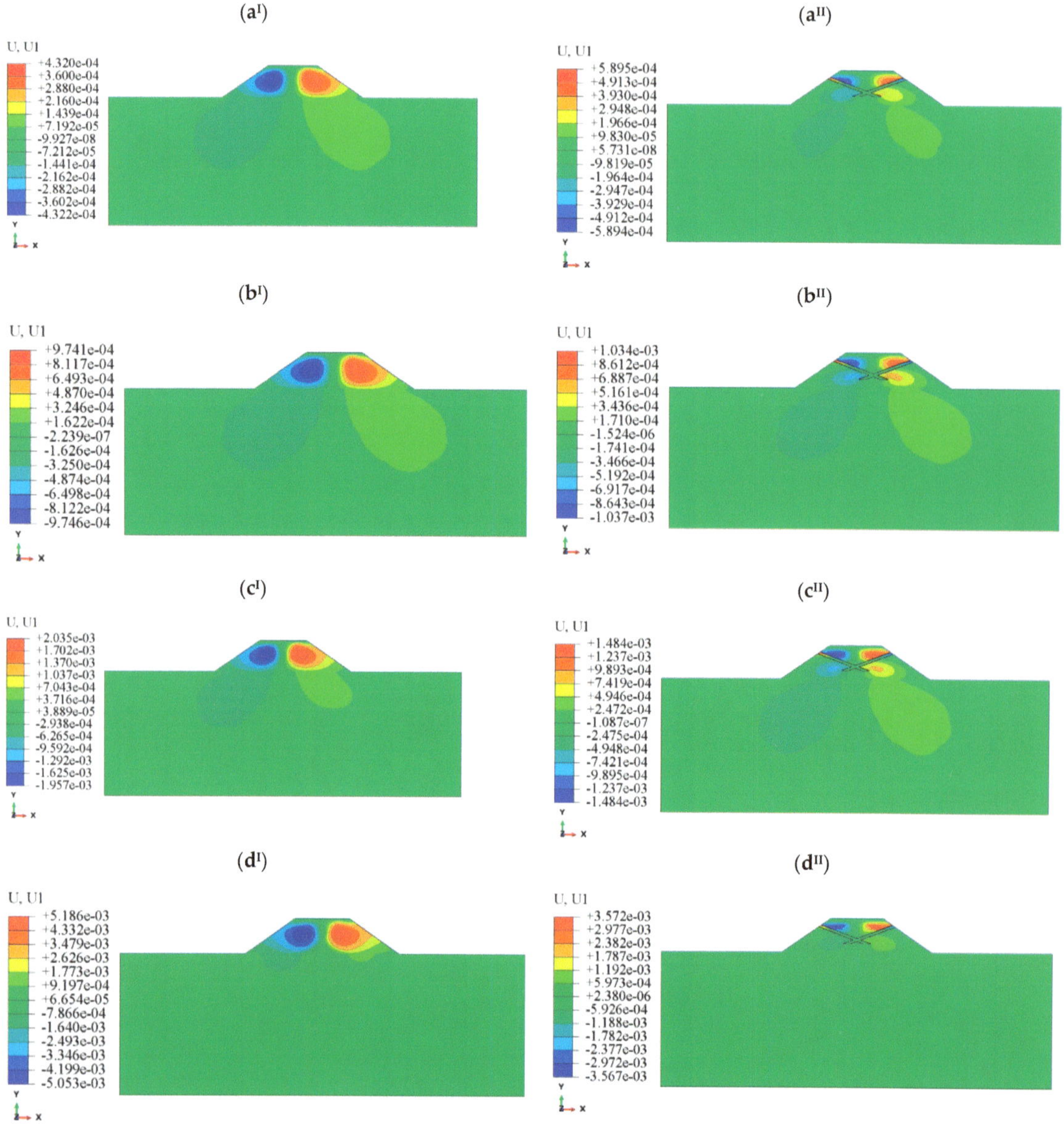

Figure 7. Cloud image of reinforcement before and after horizontal displacement (Unit: m). Unreinforced embankment. (**a**$^{\mathbf{I}}$) First month; (**b**$^{\mathbf{I}}$) second month; (**c**$^{\mathbf{I}}$) third month; and (**d**$^{\mathbf{I}}$) fourth month. Reinforced embankment. (**a**$^{\mathbf{II}}$) First month; (**b**$^{\mathbf{II}}$) second month; (**c**$^{\mathbf{II}}$) third month; and (**d**$^{\mathbf{II}}$) fourth month.

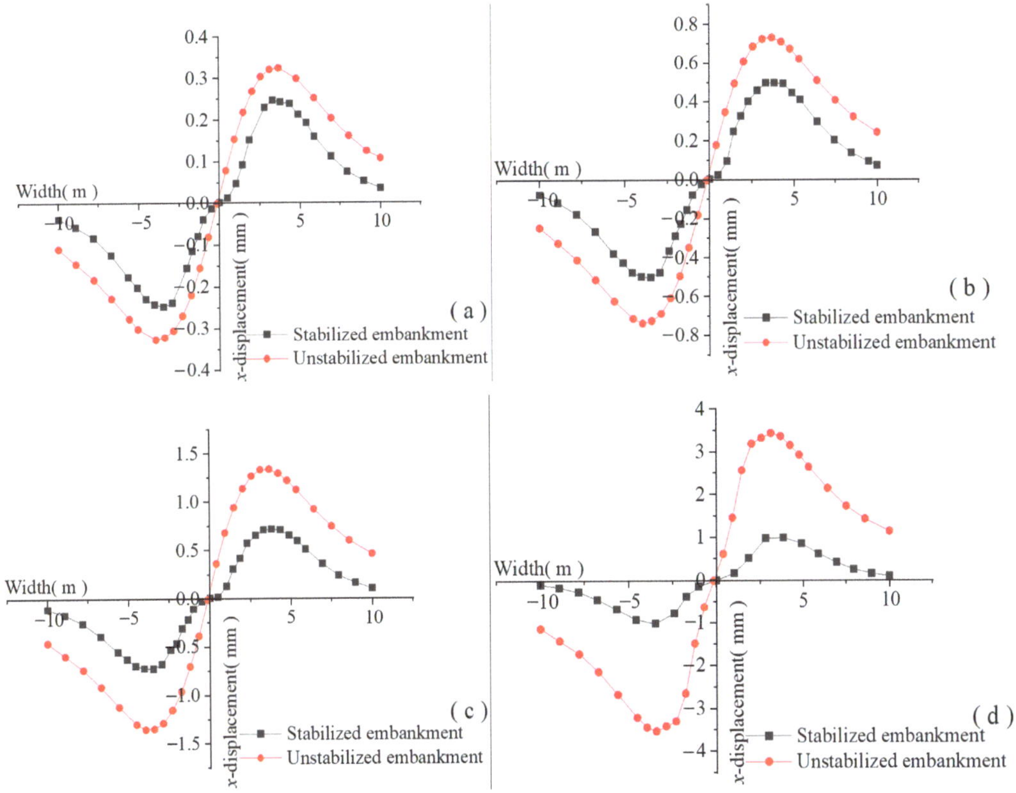

Figure 8. Monthly horizontal displacement (X-displacement) comparison of the embankment bottom before and after consolidation. (**a**) First month; (**b**) second month; (**c**) third month; and (**d**) fourth month.

5. Discussion

In this study, our results indicated that the ISCW changed the characteristics of the existing WFS embankment. Because of the ISCW support effects on the embankment, the embankment displacement decreased considerably; however, the stress variation law of the ISCW should be studied. The left and right ISCWs were arranged symmetrically. Therefore, the distribution of the displacement of the ISCW was also symmetric. Figure 9 shows the variation in pile stress along the ISCW under different pavement loads. It can be seen that the pile stress increases with the pavement load, and the stress of the piles exhibits maximum stress in the right and left ISCW intersection points.

The deformation of the ISCW is caused by the load on the embankment. The ISCW reduced the z-displacement and x-displacement from the embankment. According to the experiment results of Ye et al. [36], if the thaw collapse displacement is no more than 2 cm, the frozen soil subgrade can be considered stable. The maximum displacement with the ISCW was 1.73 cm, which occurred during the fourth month. According to these findings, the proposed technique effectively stabilized the existing WFS embankment.

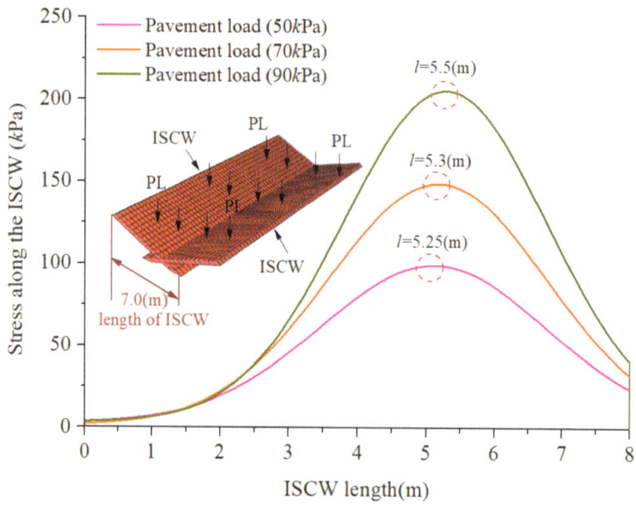

Figure 9. Effect of different pavement loads along the ISCW length.

6. Conclusions and Recommendations

According to the obtained numerical simulation results, an effective novel method was proposed for the reinforcement of WFS embankments. The main conclusions can be summarized as follows:

(1) In the cold season, the ISCW can absorb the negative temperature of the outside world and transfer it to the deeper soil layer of the embankment, decreasing the temperature of the deep, frozen soil. With an increase in the outside air temperature, the continuous wall under the ISCW can transfer the negative temperature of the deeper frozen soil to the embankment, thus reducing the temperature of the embankment. The minimum temperature of the composite embankment during the considered four-month period was 2.4 °C lower than that of the unreinforced embankment. The continuous wall under inclined cement soil can effectively reduce the embankment temperature and slow down embankment melt settlement.

(2) A composite embankment is formed between the ISCW and the soil mass, which improves the integrity of the embankment. The maximum melt settlement of the composite embankment was 28% less than that of the unreinforced embankment. The ISCW can effectively reduce the melt settlement of a frozen soil embankment at higher temperatures, thus improving the bearing capacity and stability of the embankment.

(3) The maximum horizontal displacement of the composite embankment was reduced by 30% compared with the unreinforced embankment, and the horizontal displacement at the bottom of the embankment was reduced by 67.8%. The deformation of the embankment mainly derives from the load on the road surface. As the inclined continuous wall has a "clamping" effect on the soil mass, the maximum horizontal displacement occurs in the underground continuous wall, which has a significant effect on reducing the horizontal displacement of the embankment.

Based on the melting process over four months, after comparative analysis and evaluation of the reinforcement effect of the composite embankment against that of the unreinforced embankment, it is considered that the ISCW provides a feasible approach for the reinforcement of WFS embankments. The proposed technology provides a scientific and theoretical basis for the prevention and control of existing embankment melt subsidence disasters in WFS areas. Thus, the novel method to reduce melt subsidence disasters in existing WFS embankments can be recommended to support the infrastructural integrity of major projects along the Qinghai–Tibet engineering corridor (QTEC) in the future.

Author Contributions: Conceptualization, G.S.; Methodology, L.L. and Y.H.; Software, L.L. and Y.H.; Validation, Z.T.; Investigation, Z.F., G.L. (Gao Lv) and T.C.; Resources, G.L. (Gao Lv) and T.C.; Data curation, Z.F., M.S. and G.L. (Guanqi Lan); Writing—original draft, G.S.; Writing—review & editing, G.S.; Visualization, Z.W. and Z.T.; Supervision, Z.W.; Project administration, M.S. and G.L. (Guanqi Lan). All authors have read and agreed to the published version of the manuscript.

Funding: This research was funded by [National Natural Science Foundation of China] grant number [42301156 and 52308204], [Natural Science Foundation of Shanxi Province] grant number [2023-JC-QN-0626 and 2024JC-YBMS-436], [Shaanxi Provincial Key Laboratory of Geotechnical and Underground Space Engineering open fund] grant number [YT202302] and [Xi' an Shiyou University graduate student innovation and practical ability training program] grant number [YCS23214312].

Data Availability Statement: The data presented in this study are available on request from the corresponding author.

Conflicts of Interest: Author Zonghui Fang was employed by the company Gansu Institute of Architecture Design and Research Co., Ltd. Author Ziqiang Tang was employed by the company China State Construction Engineering Corporation Aecom Consultant Co., Ltd. The remaining authors declare that the research was conducted in the absence of any commercial or financial relationships that could be construed as a potential conflict of interest.

Notation

Abbreviation	Explanation
T_s	freezing or thawing state temperature
T_f	freezing state temperature
T_u	thawing state temperature
C_f	volumetric heat capacity in the freezing state
λ_f	thermal conductivity in the freezing state
C_u	volumetric heat capacity in the thawing state
λ_u	thermal conductivity in the thawing state
T_m	freezing temperature
ΔT	temperature increment
C_s	heat capacity of frozen soil
λ_s	heat conductivity coefficient of frozen soil
σ_x, σ_y	x- and y-direction normal stresses
$\varepsilon_x, \varepsilon_y$	x- and y-direction plane shear strains
ε_{xy}	plane shear strain in the xy-plane
$E(T)$	elastic modulus of a material
μ	Poisson's ratio
α	coefficient of thermal expansion
$\{d\sigma\}$	stress increment matrix
$\{D_T\}$	elastic matrix related to temperature
$\{d\varepsilon\}$	total strain increment matrix
$\{d\varepsilon_p\}$	plastic strain increment matrix
$d\lambda$	plastic multiplier
J_2	second deviatoric stress invariant
I_1	first stress invariant
c, φ	cohesion and internal friction angle for frozen soil

References

1. Jin, H.; Jin, X.; He, R.; Luo, D.; Chang, X.; Wang, S.; Marchenko, S.S.; Yang, S.; Yi, C.; Li, S.; et al. Evolution of permafrost in China during the last 20 ka. *Sci. China Earth Sci.* **2019**, *62*, 1207–1223. [CrossRef]
2. Zou, D.; Zhao, L.; Sheng, Y.; Chen, J.; Hu, G.; Wu, T.; Wu, J.; Xie, C.; Wu, X.; Pang, Q.; et al. A new map of permafrost distribution on the Tibetan Plateau. *Cryosphere* **2017**, *11*, 2527–2542. [CrossRef]
3. Zhang, Z.; Zhang, H.; Zhang, J.; Chai, M. Effectiveness of Ionic Polymer Soil Stabilizers on Warm Frozen Soil. *KSCE J. Civ. Eng.* **2019**, *23*, 2867–2876. [CrossRef]
4. Peng, Y.; He, J.; Jiang, H. Experimental Study on Creep Mechanical Properties of Warm Frozen Soils. *Indian Geotech. J.* **2023**, *53*, 1347–1354. [CrossRef]

5. Zhang, Z.Q.; Wu, Q.B. Thermal hazards zonation and permafrost change over the Qinghai–Tibet Plateau. *Nat. Hazards* **2012**, *61*, 403–423. [CrossRef]
6. Cui, K.; Qin, X.T. Landslide risk assessment of frozen soil slope in Qinghai Tibet Plateau during spring thawing period under the coupling effect of moisture and heat. *Nat. Hazards* **2023**, *115*, 2399–2416. [CrossRef]
7. Zhao, R.; Wang, S.; Huang, X.; Ma, T.; Chen, J.; Luo, H. Research on the deformation and damage process of crushed-rock highway embankment in permafrost areas. *Comput. Part. Mech.* **2024**. [CrossRef]
8. Wu, G.; Xie, Y.; Wei, J.; Yue, X. Experimental study and prediction model on frost heave and thawing settlement deformation of subgrade soil in alpine meadow area of Qinghai-Tibet Plateau. *Arab. J. Geosci.* **2022**, *15*, 534. [CrossRef]
9. Zhang, F.; Shi, S.; Feng, D.-C.; Cai, L.-Z. Investigation on creep behavior of warm frozen silty sand under thermo−mechanical coupling loads. *J. Mt. Sci.* **2021**, *18*, 1951–1965. [CrossRef]
10. Sun, X.; Song, J.; Guo, Y.; Zhang, Y.; He, S. Analysis of thermal impact of highway construction in plateau region on frozen soil corridor. *Arab. J. Geosci.* **2021**, *14*, 1271. [CrossRef]
11. He, P.; Niu, F.; Huang, Y.; Zhang, S.; Jiao, C. Distress Characteristics in Embankment-Bridge Transition Section of the Qinghai-Tibet Railway in Permafrost Regions. *Int. J. Disaster Risk Sci.* **2023**, *14*, 680–696. [CrossRef]
12. Pan, H.; Tong, L.; Wang, Z.; Yang, T. Effects of Soil–Cement Mixing Wall Construction on Adjacent Shield Tunnel Linings in Soft Soil. *Arab. J. Sci. Eng.* **2022**, *47*, 13095–13109. [CrossRef]
13. Prakash, K.G.; Krishnamoorthy, A. Stability of Embankment Constructed on Soft Soil Treated with Soil–Cement Columns. *Transp. Infrastruct. Geotechnol.* **2022**, *10*, 595–615. [CrossRef]
14. Alzabeebee, S.; Abdulqader Alkhalidi, S.H.; Keawsawasvong, S. Numerical analysis of the dynamic response of a ballastless railway track rested on non-stabilized and cement-stabilized soft soil considering undrained conditions. *Innov. Infrastruct. Solut.* **2023**, *8*, 334. [CrossRef]
15. Maleki, M.; Khezri, A.; Nosrati, M.; Hosseini, S.M.M.M. Seismic amplification factor and dynamic response of soil-nailed walls. *Model. Earth Syst. Environ.* **2023**, *9*, 1181–1198. [CrossRef]
16. Maleki, M.; Mohammad Hosseini, S.M.M. Assessment of the Pseudo-static seismic behavior in the soil nail walls using numerical analysis. *Innov. Infrastruct. Solut.* **2022**, *7*, 262. [CrossRef]
17. Sun, G.-C.; Yao, G.; Zhang, J.-M.; Li, B.; Li, J.-Q.; Lian, W.-P.; Wei, Y. Stabilized effects of L-S cement-mixed batter pile composite foundation for existed warm frozen soil subgrade. *J. Mt. Sci.* **2023**, *20*, 542–556. [CrossRef]
18. Sun, G.-C.; Zhang, J.-M.; Dang, Y.-S.; Ding, C. Microstructure and strength features of warm and ice-rich frozen soil treated with high-performance cements. *J. Mt. Sci.* **2019**, *16*, 1470–1482. [CrossRef]
19. Shabeer, S.; Asad, S.; Jamal, A.; Ali, A. Train-induced vibration and subsidence prediction of the permafrost subgrade along the Qinghai-Tibet Railway. *Soil Dyn. Earthq. Eng.* **2022**, *14*, 307. [CrossRef]
20. Chai, M.; Zhang, J.; Ma, W.; Yin, Z.; Mu, Y.; Zhang, H. Thermal influences of stabilization on warm and ice-rich permafrost with cement: Field observation and numerical simulation. *Appl. Therm. Eng.* **2019**, *148*, 536–543. [CrossRef]
21. Tang, L.; Yang, L.; Wang, X.; Yang, G.; Ren, X.; Li, Z.; Li, G. Numerical analysis of frost heave and thawing settlement of the pile–soil system in degraded permafrost region. *Environ. Earth Sci.* **2023**, *80*, 693. [CrossRef]
22. Zhang, G.; Liu, E.; Zhang, G.; Chen, Y. Study thermo-hydro-mechanical coupling behaviors of saturated frozen soil based on granular solid hydrodynamics theory. *Arch. Appl. Mech.* **2023**, *91*, 3921–3936. [CrossRef]
23. Mu, S.; Ladanyi, B. Modelling of coupled heat, moisture and stress field in freezing soil. *Cold Reg. Sci. Technol.* **1987**, *14*, 237–246. [CrossRef]
24. Chen, W.F.; Mizuno, E. *Nonlinear Analysis in Soil Mechanics: Theory and Implementation*; Elsevier: Amsterdam, The Netherlands, 1990.
25. Sun, Z.-Z.; Ma, W.; Zhang, S.-J.; Mu, Y.-H.; Yun, H.-B.; Wang, H.-L. Characteristics of thawed interlayer and its effect on embankment settlement along the Qinghai-Tibet Railway in permafrost regions. *J. Mt. Sci.* **2018**, *15*, 1090–1100. [CrossRef]
26. Wang, T.; Liu, J.; Tai, B.; Zang, C.; Zhang, Z. Frost jacking characteristics of screw piles in seasonally frozen regions based on thermo-mechanical simulations. *Comput. Geotech* **2017**, *91*, 27–38. [CrossRef]
27. Liu, M.-H.; Li, G.-Y.; Niu, F.-J.; Lin, Z.-J.; Shang, Y.-H. Porosity of crushed rock layer and its impact on thermal regime of Qinghai−Tibet Railway embankment. *J. Cent. South Univ.* **2017**, *24*, 977–987. [CrossRef]
28. Li, K.-Q.; Yin, Z.-Y.; Qi, J.-L.; Liu, Y. State-of-the-Art Constitutive Modelling of Frozen Soils. *Arch. Comput. Methods Eng.* **2024**. [CrossRef]
29. *GB50010-2010*; Standards Press of China Code for Design of Concrete Structures. China Architecture and Building Press: Beijing, China, 2010.
30. *GB50176-93*; China Academy of Building Research Code for Thermal Design of Civil Buildings. China Planning Press: Beijing, China, 1993.
31. Chen, J.; Yuan, Y.; Yang, X.; Wang, Z.; Kang, S.; Wen, J. The Characteristics and Controlling Factors of Water and Heat Exchanges over the Alpine Wetland in the East of the Qinghai–Tibet Plateau. *Adv. Atmos. Sci.* **2016**, *40*, 201–210. [CrossRef]
32. Tang, X.J.; Wang, J. A general shakedown approach for geo-structures under cyclic loading using ABAQUS/Python. *Acta Geotech.* **2022**, *17*, 5773–5788. [CrossRef]
33. Beiranvand, B.; Rozbahani, M.Z.; Mazaheri, A.R.; Komasi, M. Quasi-static and dynamic analysis of pore water pressure in Azadi earth dams using Abaqus software. *Arab. J. Geosci.* **2021**, *14*, 1280. [CrossRef]

34. Wang, T.Y.; Yan, L.E. A heat-flux upper boundary for modeling temperature of soils under an embankment in permafrost region. *Sci. Rep.* **2022**, *12*, 13295. [CrossRef]
35. Yin, G.-A.; Niu, F.-J.; Lin, Z.-J.; Luo, J.; Liu, M.-H. Performance comparison of permafrost models in Wudaoliang Basin, Qinghai-Tibet Plateau, China. *J. Mt. Sci.* **2016**, *13*, 1162–1173. [CrossRef]
36. Yu, F.; Qi, J.; Yao, X.; Liu, Y. Degradation process of permafrost underneath embankments along Qinghai-Tibet Highway: An engineering view. *Cold Reg. Sci. Technol.* **2013**, *85*, 150–156. [CrossRef]

buildings

MDPI

Article

Evaluation of Maximum Shear Strength of Prestressed Concrete (PSC) Hollow Core Slab (HCS)

Dong-Hwan Kim, Min-Su Jo, Su-A. Lim, Hyeong-Gook Kim, Seong-Won Kang and Kil-Hee Kim *

Department of Architectural Engineering, Kongju National University, 1223-24, Cheonandaero, Seobuk, Cheonan-si 31080, Republic of Korea; kimdh@kongju.ac.kr (D.-H.K.); msjo1982@kongju.ac.kr (M.-S.J.); sualim0126@kongju.ac.kr (S.-A.L.); anthk1333@kongju.ac.kr (H.-G.K.); rkdtjd1652@kongju.ac.kr (S.-W.K.)
* Correspondence: kimkh@kongju.ac.kr; Tel.: +82-41-521-9335

Abstract: In this study, four-point load tests were conducted to evaluate the shear performance of factory-produced precast prestressed concrete hollow core slabs (HCS) assembled on-site. The test specimens were fabricated using compression molding and comprised six samples, with variables being the presence or absence of topping concrete and the shear reinforcement. According to the experimental variables, experiments were conducted using simple support beams to evaluate the shear performance and ultimate strength of HCSs. The results showed that HCSs, regardless of whether they included topping concrete or not, exhibited average values of shear strength more than 10% higher than the factored shear strength specified by concrete structure standards, confirming that these materials satisfy existing design standards. According to current standards, the overall reinforcement length should be increased to meet the minimum shear rebar placement requirements. However, the nominal shear strength of PS concrete hollow slabs exceeded the hollow design, with the ratio of experimental results ranging from 1.26 to 1.87 on average, satisfying the required performance.

Keywords: shear strength; hollow core slabs; precast concrete; prestressed concrete

Citation: Kim, D.-H.; Jo, M.-S.; Lim, S.-A.; Kim, H.-G.; Kang, S.-W.; Kim, K.-H. Evaluation of Maximum Shear Strength of Prestressed Concrete (PSC) Hollow Core Slab (HCS). *Buildings* **2024**, *14*, 1925. https://doi.org/10.3390/buildings14071925

Academic Editors: Grzegorz Ludwik Golewski and Yin Chi

Received: 24 May 2024
Revised: 13 June 2024
Accepted: 20 June 2024
Published: 24 June 2024

1. Introduction

Prestressed concrete hollow core slabs (PS-HCS) are structurally optimized precast slab members designed to reduce the amount of concrete material and the self-weight by forming elliptical or polygonal hollow cavities in the slab. By introducing prestress at the top, bottom, and other required locations, these slabs achieve significantly improved stiffness compared to conventional reinforced concrete slabs. Since PS-HCS members are manufactured in precast form, they can be mass-produced in factories, ensuring much better quality than general cast-in-place concrete members. Additionally, PS-HCSs offer excellent economic advantages, such as reduced temporary construction costs during on-site installation, reduced construction time, and lower labor costs due to quick assembly. Moreover, HCSs minimize construction waste and enhance material efficiency for concrete and steel, making them an eco-friendly construction method [1–7].

HCSs were first developed in the 1950s and have since been used in various structural applications [8]. PS-HCSs, with top and bottom tension members, are widely used throughout North America and Europe due to their superior bending performance compared to that of reinforced concrete slabs [9–12]. Because HCSs are produced in factories and assembled on-site, they can be actively applied to architectural structures requiring long-span floors, such as parking lots, logistics centers, discount stores, and semiconductor factories. Their use reduces construction time and facilitates site management compared to general reinforced concrete structures; the reduction in self-weight due to hollowing is especially advantageous for long-span structures [13]. However, since HCSs are extruded by compressing and extruding high-strength concrete with low slump and without formwork, embedding shear reinforcement rods during manufacturing is almost impossible [14].

Therefore, at ends, where shear force is dominant, a structural review of shear strength is crucial due to stress concentration in the concrete caused by the introduction of prestress.

Previous research has found that the evaluation of the horizontal shear strength in pushover tests, with the manufacturing method and interface roughness as variables, is significantly affected by interface roughness, orientation, and the presence of laitance [15]. Researchers also compared prior work on shear friction and analyzed key factors influencing the interface, highlighting the need to consider shrinkage and stiffness at the topping concrete–concrete surface [16]. Most researchers have evaluated the shear strength of HCSs with topping concrete in terms of interface roughness. However, there is a lack of research on the shear behavior, including the ultimate shear capacity, of PS-HCSs with composite sections that include poured topping concrete. Compared to similar reinforced concrete slabs, PS-HCSs can lead to a 20–40% reduction in material [17,18]. They are typically designed to resist bending moments under equally distributed loads and are widely used in floor decks of office buildings, residential buildings, and parking structures. However, large, concentrated loads can cause shear failure in the web section. During the manufacturing process of prestressed concrete hollow core slabs, placing shear reinforcement is difficult, so shear stress is generally resisted by the concrete's shear strength. Traditional methods involve planning deeper slabs or filling slab cores with concrete, but this reduces shear strength due to the size effect. Thus, research and development are needed to prevent shear failure while maintaining the advantages of PS-HCS.

The current ACI 318 and KDS 14 20 22 standards [19,20] specify that for HCS members with slab thicknesses greater than 315 mm and without minimum shear reinforcement, the factored load acting on the slab must not exceed half of the calculated abdominal shear capacity. Given the difficulty of reinforcing PS-HCS members against shear, the use of PS-HCSs with slab thicknesses greater than 315 mm may be limited by their shear capacity.

Therefore, this study aims to evaluate the shear performance of full-size PS-HCSs with section heights (h) of 350 mm. Shear reinforcement rods were placed at the ends to satisfy the shear reinforcement required by current standards. Shear experiments were conducted with the presence and absence of topping concrete and shear reinforcement as variables. The required structural performance was experimentally verified by comparing the results predicted by design criteria with the experimental outcomes.

2. Major Structural Design Criteria of PS-HCSs

Shear Strength

The design criteria for ACI 318-19 and shear and torsion of concrete structures (KDS 14 20 22) require that the shear strength of prestressed concrete members be at least equal to the factored shear strength, as shown in Equation (1).

$$V_u \le \phi V_n \tag{1}$$

Here, V_u is the factored shear strength, and V_n is the nominal shear strength.

The shear strength of HCS members without shear reinforcement is determined by two types of shear failure, as shown in Equations (2) and (3): shear cracking concrete shear strength (V_{ci}) and abdominal shear cracking concrete shear strength (V_{cw}). The standards require using the smaller value between these two types.

$$V_{ci} = 0.05\lambda\sqrt{f_{ck}}b_w d_p + V_d + \frac{V_i M_{cre}}{M_{max}} \ge 0.17\lambda\sqrt{f_{ck}}b_w d_p \tag{2}$$

$$V_{cw} = \left(0.29\lambda\sqrt{f_{ck}} + 0.3f_{pc}\right)b_w d_p + V_p \tag{3}$$

where λ is the lightweight concrete coefficient, f_{ck} is the maximum compressive strength of the concrete, b_w is the web width, d_p is the distance from the compression edge to the center of the tension member's cross-section, which should not be less than 0.8 h, V_d is the shear force caused by dead load, V_i is the factored shear force due to the applied load, M_{cre}

is the cracking moment due to the applied load, M_{max} is the maximum factored moment due to the applied load, f_{pc} is the compressive stress in the concrete at the centroid of the section resisting the applied load, considering the stress losses in prestress, and V_p is the vertical component of the prestressing force when inclined tensioning is used.

For composite HCSs, where topping concrete is poured on top of PC HCS, the high-strength PC concrete and the relatively low-strength CIP concrete act together. In this case, the design criteria do not provide a clear design method, so the shear strength evaluation method presented by PCI (PCI 2017) [21] can be followed. The PCI method uses the elastic modulus ratio of the topping concrete to calculate the equivalent section. Then, the neutral axis location is determined for the equivalent section, and the shear strength is calculated based on the principal stress at that location. It is assumed that the topping concrete in the area where the HCS web is extended is integrated with the same strength as the HCS, and the web shear strength can be calculated as shown in Equation (4).

$$V_n = \left(0.29\lambda\sqrt{f_{ck(HCS)}} + 0.3f_{pc}\right)b_w d_{com} \quad (4)$$

where d_{com} is the distance (mm) from the compression edge of the composite section to the center of the tension material.

In HCSs, the shear strength is determined by the web shear cracking concrete shear strength because the web, due to hollowing, is the thinnest part. Since it is crucial to calculate a reasonably effective section for an HCS when evaluating shear strength, PCI uses only the web width to calculate the design section force.

HCSs should be used to prevent web shear cracking in spans in which the effective prestress does not sufficiently develop. The web shear cracking concrete shear strength derived from Equation (3) tends to decrease by about 30% when the reduction in effective prestress at the critical section is accounted for. Furthermore, Equation (3) is based on a previous experiment [22], which pointed out that HCS members taller than 380 mm show results of web shear cracking concrete shear strength in a range of 58% to 85%, indicating that applying Equation (3) to HCS members taller than 300 mm may overestimate the actual strength. Therefore, if minimum shear reinforcement is not provided, KDS 14 20 22 and ACI 318-19 require a minimum amount of shear reinforcement for heights exceeding 315 mm in order to reduce the web shear cracking concrete shear strength by half. KDS 14 20 22 recommend minimum shear reinforcement because high net cross-sectional heights in HCSs may overestimate the design strength. While the shear strength design for HCSs combines the contributions of concrete and shear reinforcement, the actual behavior may not align with the design assumptions. As can be seen in Figure 1, the locations for calculating V_c and V_s differ between the PC web and the CIP-filled cavity. For this reason, their arithmetic summation requires the PC web concrete and shear reinforcement to reach their maximum strengths simultaneously, necessitating fully composite behavior. However, since shear reinforcement is embedded throughout the CIP, it may be difficult to fully account for the contribution of shear reinforcement if the fill concrete does not act in unison with the PC HCS.

An HCS is used in composite sections by pouring topping concrete on-site to ensure slab continuity according to structural application. For composite HCS members, the method suggested by PCI can be used to calculate the web shear cracking concrete shear strength after replacing the topping concrete with an equivalent section. Generally, the compressive strength of topping concrete poured on-site is lower than that of an HCS member's concrete, so the elastic modulus ratio of concrete ($n = E_{RC}/E_{HCS} \leq 1$) is used to calculate the web shear cracking concrete shear strength.

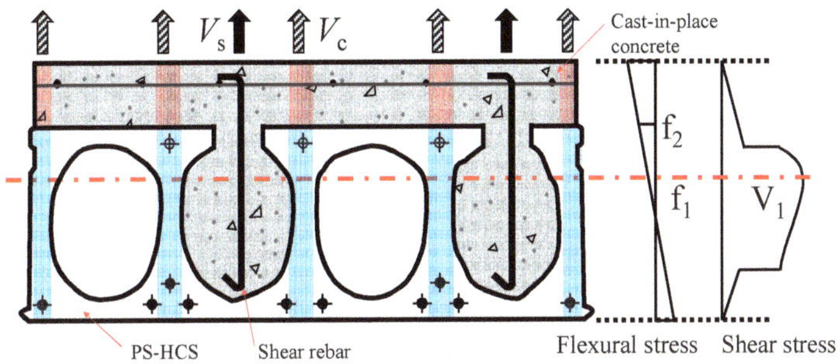

Notation: V_c represents the shear strength provided by the concrete itself. V_s represents the shear strength provided by the shear reinforcement. Additionally, it is important to consider the flexural stress f_1 without topping, the flexural stress f_2 with topping, and the shear stress V_1.

Figure 1. Web shear strength of the composite hollow core slab.

3. Test Program

3.1. Materials

The material test results shown in Table 1 were determined to evaluate the maximum shear strength of prestressed hollow core slabs (PS-HCS). The design maximum compressive strength for the concrete was planned to be 49 MPa for PS-HCS and 27 MPa for cast-in-place (CIP) concrete. To ensure reliable test results, the tests were conducted in accordance with international standards ISO 1920-3:2019 and ISO 1920-4:2020 [23,24].

Table 1. Concrete compressive strength test results.

Hollow Core Slab (N/mm)^2			Topping Concrete (N/mm)^2			Note
Design	Test	Ave	Design	Test	Ave	
49	60.21	61.57	27.0	30.20	30.51	Cylinder (φ100 × 200)
	60.56			30.10		
	63.96			31.24		

Table 1 presents the compressive strength test results of hollow core slabs and topping concrete, measured in N/mm^2. For the hollow core slabs, the design strength was specified at 49 N/mm^2. The test results showed compressive strengths of 60.21 N/mm^2, 60.56 N/mm^2, and 63.96 N/mm^2, yielding an average of 61.57 N/mm^2. Similarly, for the topping concrete, the design strength was 27.0 N/mm^2. The compressive strength test results were 30.20 N/mm^2, 30.10 N/mm^2, and 31.24 N/mm^2, with an average of 30.51 N/mm^2.

To evaluate the compressive strength of the concrete, three cylindrical specimens with dimensions φ100 mm × 200 mm were fabricated. The maximum compressive strength evaluation of the specimens was conducted immediately after the PS-HCS test. The stress–strain relationship of the concrete, depicted in Figure 2, was measured using concrete wire strain gauges attached to the longitudinal and transverse sides of the specimen during the maximum compressive strength test.

The prestressing steel samples used to evaluate the maximum shear capacity of PS-HCS were SWPC 7B with diameters of φ12.7 mm and φ9.5 mm. The physical properties included an elastic modulus (E_{ps}) of 196 GPa, a yield strength (f_{py}) of 1581 MPa, a yield strain (ε_{py}) of 0.0062, a tensile strength (f_{pu}) of 1860 MPa, and a tensile strain (ε_{pu}) of 0.0800. The stress–strain relationship of the prestressing steel is shown in Figure 3 [25].

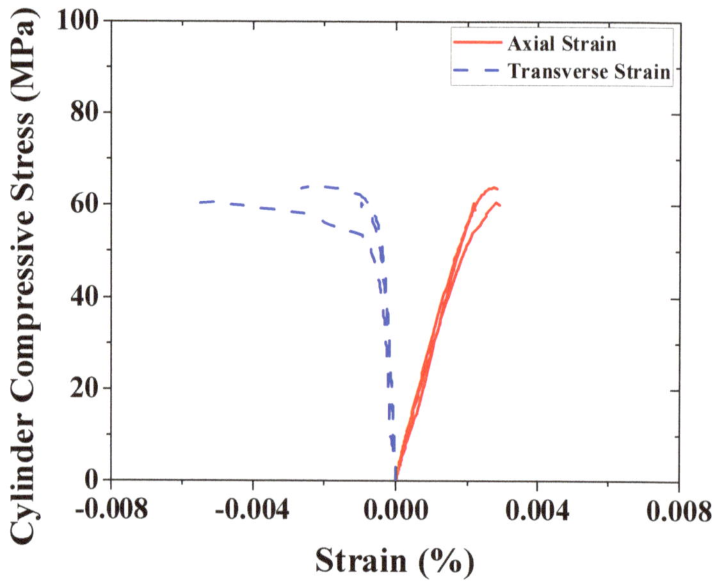

Figure 2. Relationship of concrete strain and stress.

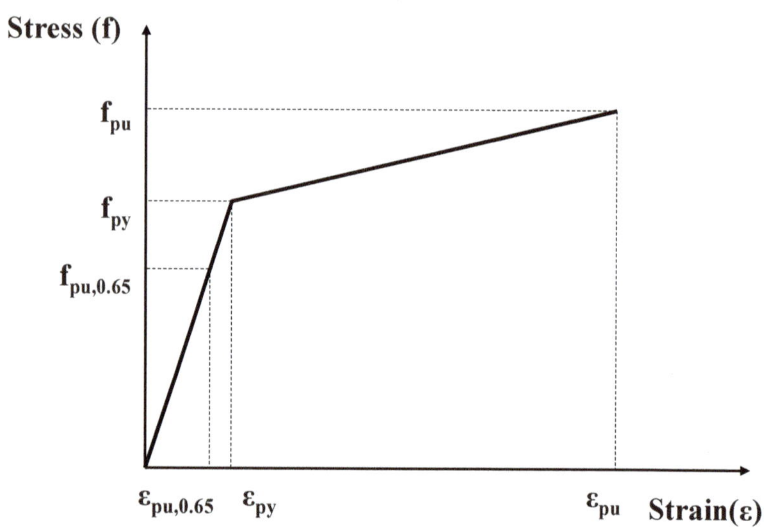

Figure 3. Relationship of tendon strain and stress [25].

The results confirm that both the PS-HCS and CIP concrete exceeded their design compressive strengths, meeting all required performance criteria.

3.2. Design of Specimens

The specimens were selected to evaluate the maximum shear performance of PS-HCS with or without overlay concrete and shear reinforcement. There were six specimens in total, with a slab cross-sectional height of 350 mm and a width of 1200 mm. All specimens had the same length of 6000 mm. The shear span ratio (a/d) of the specimens was set at 3.5 for the unreinforced specimens and 2.5 for the reinforced specimens, depending on the effective depth of the members, ensuring that shear force was dominant.

In the specimen designations shown in Figure 4 and Table 2, S refers to the shear specimen, N refers to the unreinforced specimen, and R refers to the reinforced specimen. To plan the integration of the HCS, the pouring of the overlay concrete was performed with a rough finish on the upper surface of the specimen. The prestressing steel used to fabricate the specimens was SWPC 7BL, with 10-φ12.7 at the bottom and 2-φ9.5 at the top.

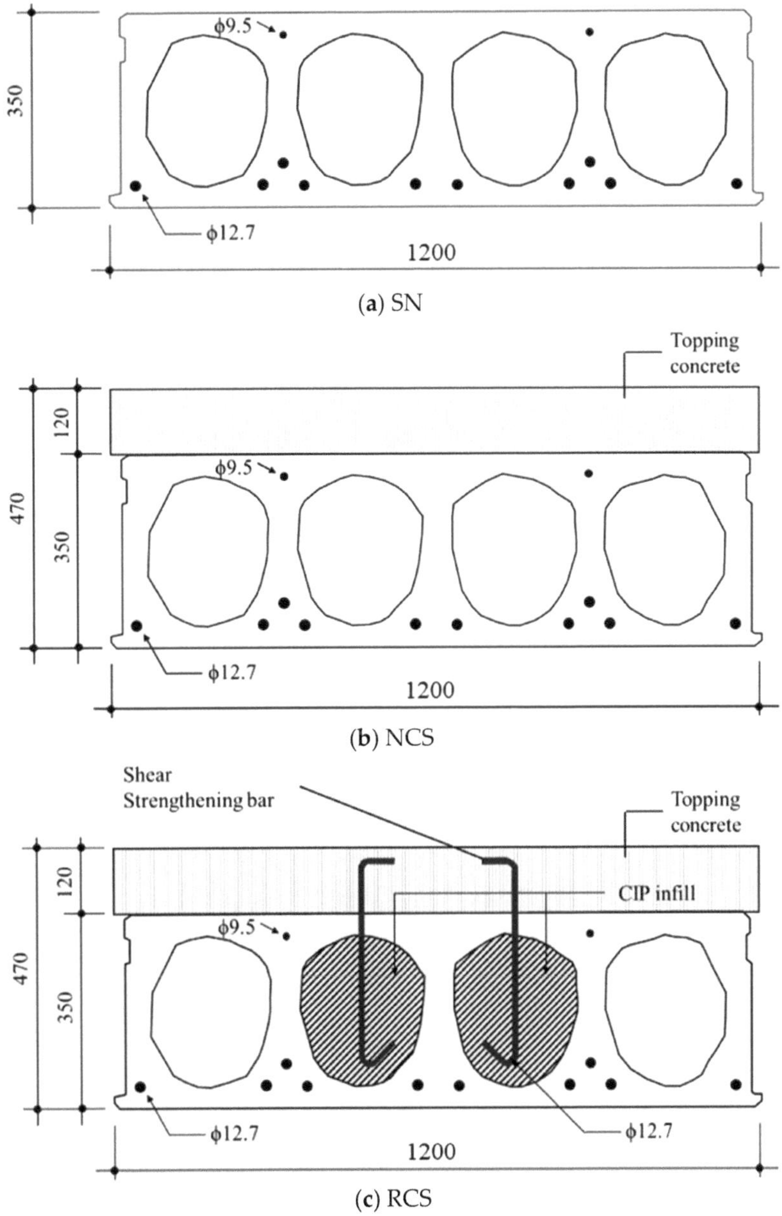

Figure 4. Detail of specimens (unit: mm).

Table 2. Design of specimens.

Specimens	f'_c (MPa)		Size (mm)			Prestressing Strand		a/d (%)	H (mm)
	HCS	Topping	*h*	*b*	*l*	Bottom	Top		
SN-1	57.6	-	350	1200	6000	10ϕ12.7	2ϕ9.5	3.5	350
SN-2									
NCS-1		30.5	470					2.5	470
RCS-1									
RCS-2									
RCS-3									

3.3. Test Setup

The specimen testing was conducted using a four-point bending test configured as a simply supported beam. This was performed with a universal testing machine (U.T.M) with a capacity of 3000 kN, as illustrated in Figure 5a. The load was applied continuously at a rate of 0.05 mm/s using the displacement control method. For the bending performance evaluation, the test continued until the load decreased after reaching its maximum shear strength. The specimen's deflection under the applied force was recorded by placing two LVDTs at the bottom of the central section between the load application points, where the maximum deformation occurred, as shown in Figure 5b.

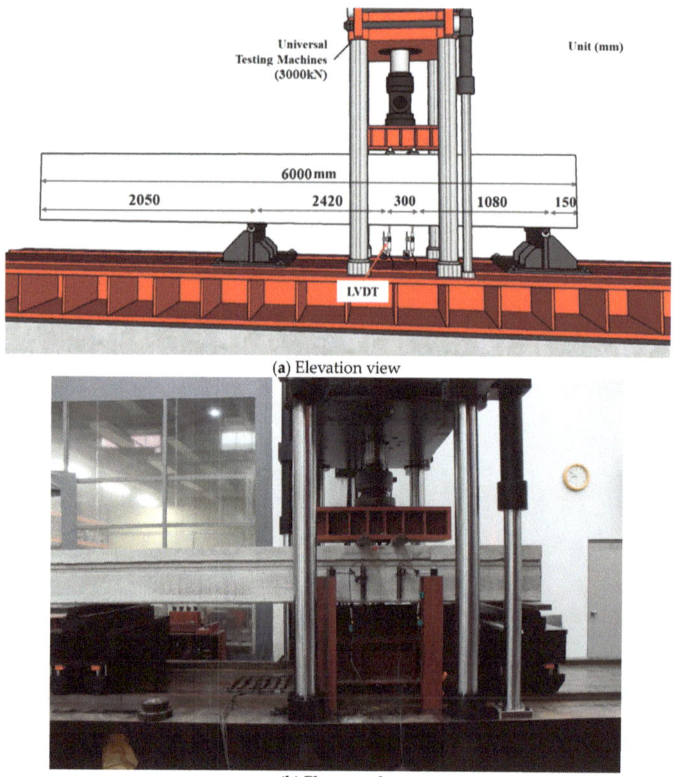

(**a**) Elevation view

(**b**) Photograph

Figure 5. Test setup (unit: mm).

Strain gauges were attached at the top of the concrete and prestressing steel of the specimen between the load-bearing points and at a point 30 mm below the top; the strain of the PS steel and concrete was measured using the fit condition of strain in a one-way slab.

4. Results and Discussion

4.1. Load–Displacement Relationship

The load–displacement relationships of the PS-HCS specimens are shown in Figures 6 and 7. The load for each specimen was measured using a load cell attached to the UTM; the deflection was measured using the displacement of an LVDT installed at the bottom center of the specimen. All specimens exhibited linear load-deflection behavior until the initial bending crack occurred. For the SN and NCS specimens without shear reinforcement, linear behavior was observed as the shear force increased after the bending crack occurred. However, after reaching the maximum strength, a diagonal shear crack appeared, causing a rapid decrease in strength. For the RCS specimens with shear reinforcement, the bending stiffness decreased as the shear force increased after the occurrence of bending cracks, similar to the SN and NCS specimens. However, the load was transferred to the shear reinforcement due to the diagonal shear cracks, resulting in a stable increase in load without a sharp decrease in strength. Eventually, the diagonal shear cracks widened significantly, leading to final failure with a decrease in strength.

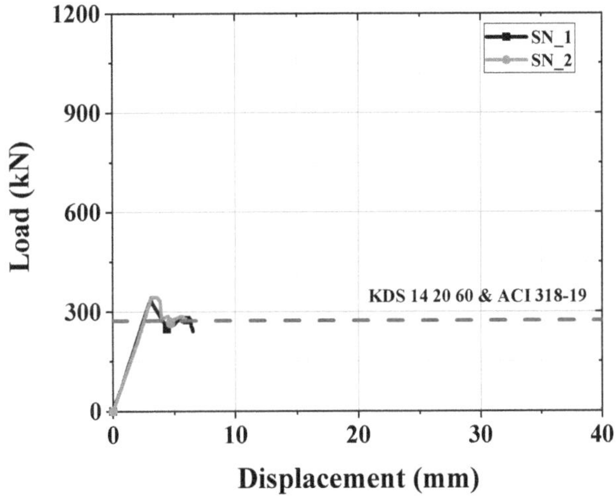

Figure 6. Load–displacement relationship of FN-Series.

The experimental results are summarized in Table 3. For the SN specimens with non-composite cross-sections, the maximum shear force was reached at an average of 343.2 kN, coinciding with the occurrence of shear cracks. The deflection at this point averaged 3.2 mm. For the composite cross-sections, the NCS specimens without shear reinforcement showed initial bending cracks at an average of 583.1 kN and a deflection of 2.6 mm, unlike the SN specimens. After the initial bending crack, the maximum shear force was reached at 625.2 kN, coinciding with the occurrence of diagonal shear cracks; the deflection was 3.8 mm. For the RCS specimens with shear reinforcement, similar to the NCS specimens, the initial bending crack occurred at an average of 573.1 kN, with an average deflection of 3.2 mm; the maximum shear capacity was reached at an average of 950.7 kN, with an average deflection of 18.2 mm.

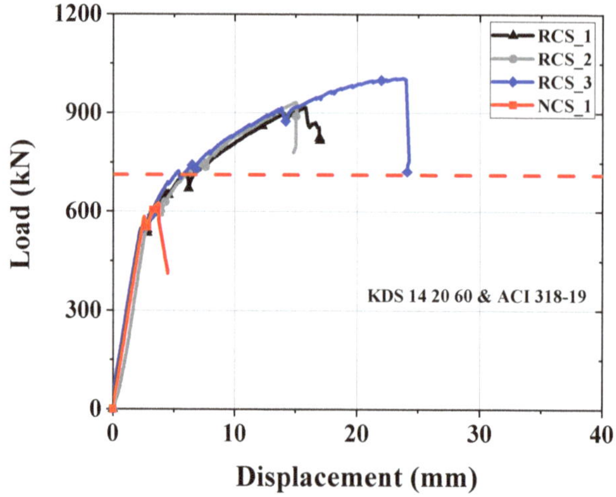

Figure 7. Load–displacement relationship of RCF-Series.

Table 3. Test result and comparison of maximum shear strength.

Specimens	Experimental Results				Analytical Results		Exp./Ana.	
	V_{Cr} (kN)	δ_{Cr} (mm)	V_{Peak} (kN)	δ_{Peak} (mm)	V_{Cr} (kN)	V_{Peak} (kN)	V_{Cr}	V_{Peak}
SN-1	177.0	11.4	322.7	120.0	181.3	317.4	0.98	1.02
SN-2	176.3	10.4	318.5	101.5			0.97	1.00
NCS-1	184.4	10.8	330.2	100.7	277.9	426.0	1.02	1.04
RCS-1	216.9	7.1	451.4	80.1			0.78	1.06
RCS-2	241.4	7.0	453.8	75.1			0.87	1.07
RCS-3	229.7	7.0	449.8	78.4			0.83	1.06

4.2. Crack Pattern

The crack patterns of the PS-HCS specimens at the maximum load are shown in Figure 8. Diagonal cracks occurred in the test section of the specimen at initial loading, regardless of reinforcement, and bending cracks were observed in the center of the test specimen at initial loading due to the lower load required for bending crack generation than for abdominal shear crack generation in the RCS specimens. The PS-HCS specimens showed a tendency in which the bending cracks in the test section propagated with increasing load; ultimately, specimens experienced shear failure due to abdominal shear cracks that occurred before the maximum load capacity was reached due to bending. The experimental results indicated that the crack angles were between 30 and 45 degrees.

The unreinforced SN specimen failed due to the occurrence of diagonal cracks in the test section at initial loading, while the RCS specimen reinforced with shear reinforcing bars showed growth of bending cracks in the pure bending section between the loading points as the load increased. The non-composite specimen failed due to the development of diagonal cracks; the composite specimen was characterized by the development of diagonal cracks, followed by an increase in strength and, finally, failure due to shear cracks in the web.

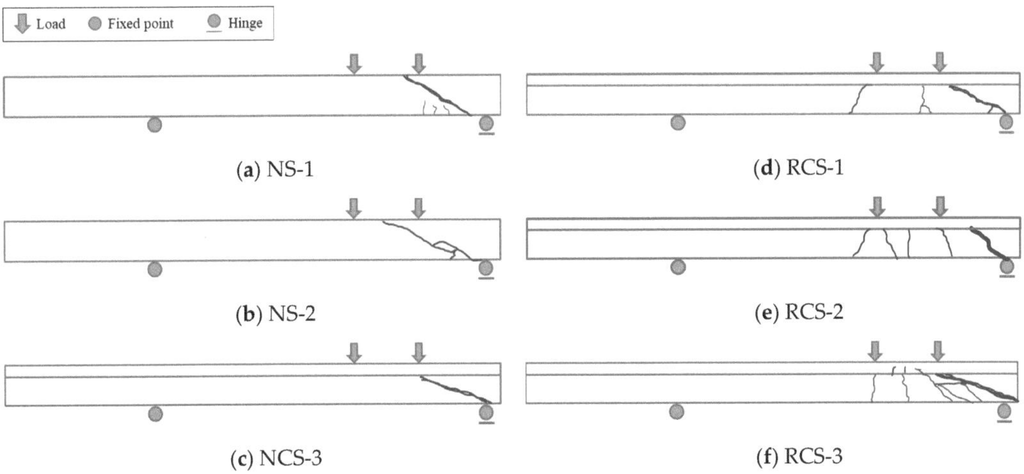

Figure 8. Shear crack patterns of PS-HCS.

4.3. Shear Strength

The shear strength (V_n) proposed by KDS 14 20 60 and ACI 318-19 was predicted from the experimental results, and the results were analyzed. The concrete shear strength of prestressed concrete members is the smaller of the bending shear cracking concrete shear strength and the web shear cracking concrete shear strength, as mentioned in Section 2. The shear strength of the PS concrete hollow slab (HCS) specimen was obtained using the load cell attached to the UTM; the predicted shear strength of the specimen was obtained using Equation (4), shown in Section 2. Additionally, the shear strength of the PS-HCS specimen was calculated using only the width of the web, as suggested by the Precast Concrete Institute (PCI).

Table 3 shows a comparison of experimental results for nominal shear strength (V_n) of PS concrete hollow slab (HCS) specimens. As can be seen, the SN specimens with non-composite cross-sections had an average shear strength of 343.2 kN and a nominal shear strength (V_n) of 273.2 kN, indicating that the ratio of the experimental results was, on average, more than 1.26. For the NCS specimens with composite cross-sections, the average shear strength was 625.2 kN, and the nominal shear strength was 334.5 kN, resulting in a ratio 1.87 times higher. RCS specimens with shear reinforcement showed an average shear strength of 950.7 kN and a nominal shear strength of 711.7 kN, with an average ratio 1.34 times higher than that of the experimental results.

As mentioned in Section 2, current design criteria require that the shear strength of concrete in a hollow slab be divided into Vci, which considers the effect of bending, and Vcw, which considers the shear strength in the web only, with the shear strength Vc being the minimum of these two values. Additionally, if shear reinforcement is provided in the composite HCS, the same method is applied as the shear reinforcement design formula for flexural members.

Therefore, as shown in Table 3, the shear strength of the test specimens is 1.26 to 1.87 times that of KDS 14 20 60 without considering the strength reduction factor, sufficiently satisfying the required performance for shear strength. As mentioned in Section 2, the minimum shear reinforcement placement in KDS 14 20 22 is 1.7 to 1.8 m of the shear reinforcement at the end of a 350 mm section. In this test, 1.0 m of shear reinforcement was placed at the end, and since the performance exceeded the nominal shear strength, it is concluded that 1.0 m of shear reinforcement will provide the required structural performance.

4.4. Analysis of Influence Factors on the Web-Shear Capacity

As mentioned in Section 2, the shear performance of HCS is determined by the web shear cracking concrete shear strength rather than the bending shear cracking concrete shear strength due to the introduction of prestress and the hollowing effect in the web, which has a relatively weak structure. To objectively evaluate this, the main influencing factors were analyzed using the experimental results of previous researchers [6,14–19]. For comparison, 170 experimental data points were collected; these are summarized in Table 4. Figure 9 provides a comparison between the results of previous studies and the experimental results of HCS, with the shear strength normalized (v_{test}/f_c') for objective comparison of each major influence factor. Figure 9a,b illustrates the trend of the concrete compressive strength and the shear span ratio, indicating an increasing trend with a consistent slope as both factors increase. Figure 9c, which shows the concrete compressive stress while considering the stress loss of prestress at the center of the cross-section, reveals a significant increase in the concrete compressive stress compared to other variables, confirming the major impact of this variable on the web shear zone concrete shear strength. Figure 9d shows that, for member height, both HCS experimental results and previous studies exhibit a similar decreasing trend compared to other variables. However, when compared to the shear strength required for web shear zone concrete ($0.29fc$) in the standards, previous studies displayed a consistent trend, confirming that the size effect did not appear in HCS members, similar to the findings of Palmer and Schultz [25].

Table 4. Summary of previous studies.

Researchers	f'_c (MPa)	h (mm)	b_w (mm)	d_p (mm)	A_g (cm)^2	A_{ps} (mm)^2	a/d	f_{pc} (MPa)	V_{test} (kN)
Bertagnoli and Mancini [2] (14 specimens)	55.0 ~65.7	163.0 ~421.5	335.0 ~444.0	133.0 ~378.5	1120.0 ~2380.0	278.8 ~1435.0	2.8 ~4.5	2.3 ~6.8	97.0 ~478.0
Walraven and Mercx [4] (19 specimens)	64.0	255.0 ~300.0	250.0 ~294.0	225.0 ~265.0	1710.0 ~1990.0	470.0 ~940.0	1.7 ~6.7	2.5 ~5.9	181.6 ~286.3
Pajari [7] (50 specimens)	38.1 ~63.7	200.0 ~500.0	215.0 ~335.0	160.0 ~453.0	1190.0 ~3000.0	372.0 ~1953.0	2.7 ~5.7	1.8 ~7.0	80.4 ~528.0
Lee et al. [16]	48.2 ~28.7	265.0 ~400.0	252.0 ~275.0	158.0 ~214.0	1625 ~2055	162.5 ~216.3	1.5 ~2.6	4.0 ~5.5	295.1 ~815.5
Palmer and Schultz [25] (24 specimens)	53.9 ~68.7	304.8 ~508.0	299.5 ~438.9	256.5 ~448.8	-	954.8 ~1587.1	2.5 ~4.0	1.0 ~5.5	221.2 ~609.7
TNO [26] (39 specimens)	55.9 ~113.9	255.0 ~400.0	241.0 ~449.0	202.0 ~350.0	1720.0 ~2610.0	853.7 ~1684.4	2.9 ~3.2	3.6 ~8.8	224.0 ~652.0
Celal [27] (8 specimens)	62.9 ~67.9	206.0 ~305.0	229.0 ~313.0	158.0 ~255.0	1501.4 ~2069.9	502.5 ~888.3	3.0 ~3.8	3.6 ~5.0	163.0 ~297.0
Park et al. [28] (10 specimens)	60.5	200.0 ~500.0	242.0 ~300.0	168.0 ~452.0	1215.9 ~2554.1	504.8 ~899.6	2.5 ~3.1	4.0 ~5.1	81.9 ~454.4

A_g—cross-sectional area; A_{ps}—cross-sectional area of tendon; f_{pc}—prestress introduced in the centroid of section.

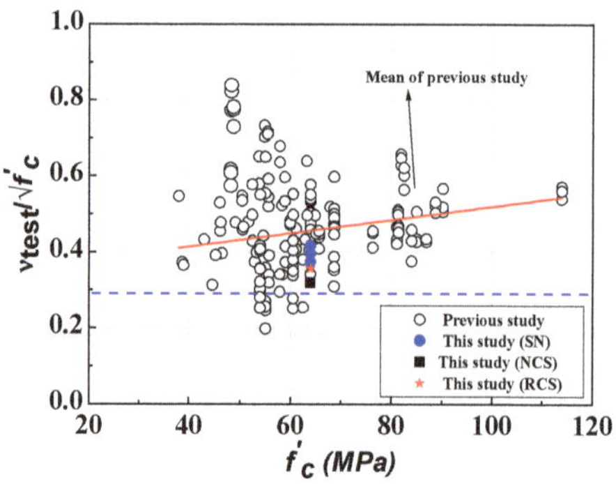

(a) Concrete compressive strength

Figure 9. *Cont.*

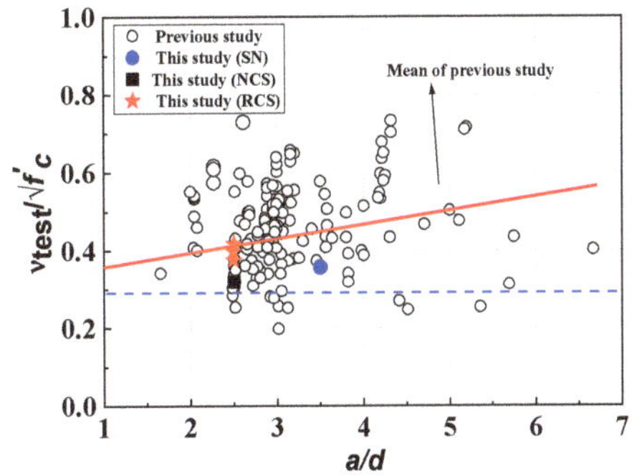

(b) Shear span-to-depth ratio

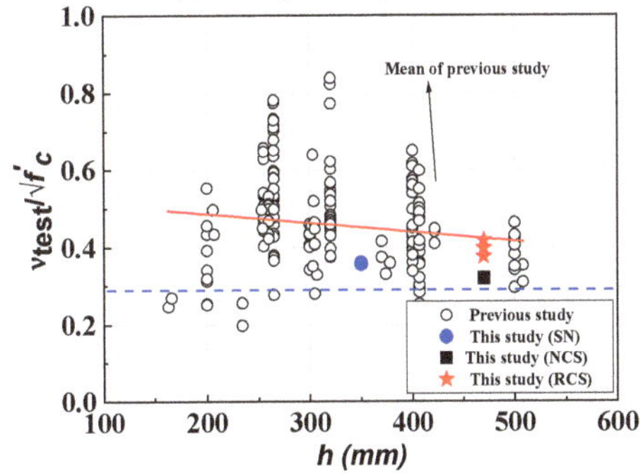

(c) Height of members

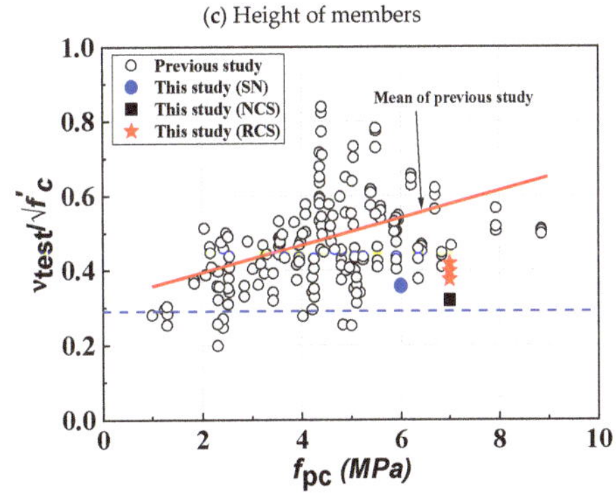

(d) Prestress introduced in centroid section

Figure 9. Analysis of key influence factors on the web-shear capacity: (**a**) Concrete compressive strength; (**b**) shear span-to-depth ratio; (**c**) Height of members; (**d**) Prestress introduced in centroid of section.

5. Conclusions

In this study, the maximum shear capacity of PS concrete hollow slabs (HCS) produced by extrusion molding was evaluated, and the following conclusions were drawn:

1. The PS concrete hollow slab (HCS) specimens with shear reinforcement exhibited increased strength and ductile behavior compared to specimens without shear reinforcement. After the occurrence of diagonal shear cracks, the shear reinforcement effectively resisted these cracks. For the unreinforced N-series specimens, the ratios of the nominal shear strength to peak shear strength were 1.02, 1.00, and 1.04, respectively. In contrast, for the reinforced RCS specimens, the ratios were 1.06,

1.07, and 1.06, demonstrating the enhanced strength and ductile behavior due to the shear reinforcement.

2. The nominal shear strength of the PS concrete hollow slab exceeded the design values considering the hollow section, with experimental results showing ratios ranging from 1.26 to 1.87. This indicates that the required performance was met, largely due to the effects of shear reinforcement, topping concrete, and the compressive stress induced by prestressing at the section's centroid, resulting in improved web shear performance.
3. According to the design criteria, the shear reinforcement length should be increased However, even the RCS specimens that did not meet the minimum shear reinforcement criteria exceeded the nominal shear strength. Therefore, the expected shear performance degradation for hollow slabs thicker than 315 mm, as restricted by the standard, was not observed in this experiment. The 350 mm section hollow slab verified in this study satisfied the current shear performance standards despite not meeting the minimum shear reinforcement criteria. While the concrete compressive strength and the shear span-to-depth ratio had some influence on web shear strength, the effect of size was found to be minimal.

Author Contributions: Conceptualization, D.-H.K. and K.-H.K.; data curation, M.-S.J. and S.-A.L.; formal analysis, D.-H.K. and S.-W.K.; funding acquisition, K.-H.K.; investigation, D.-H.K.; methodology, H.-G.K.; project administration, K.-H.K. and D.-H.K.; supervision, K.-H.K.; visualization, M.-S.J.; writing—original draft, D.-H.K.; writing—review and editing, K.-H.K. and H.-G.K. All authors have read and agreed to the published version of the manuscript.

Funding: This research was supported by the research grant of Kongju National University in 2024 and the Basic Science Research Program through the National Research Foundation of Korea (NRF) funded by the Ministry of Education (2019R1A6A1A03032988 and 2023R1A2C3002443). This research was supported by the Korea Basic Science Institute (National Research Facilities and Equipment Center) grant funded by the Ministry of Education (2022R1A6C101A741).

Data Availability Statement: The data presented in this study are available upon request from the corresponding author.

Acknowledgments: This research was conducted with experimental support from Hansung PCC.

Conflicts of Interest: The authors declare no conflicts of interest. The funders had no role in the study design, collection, analysis, and interpretation of data, the writing of the manuscript, or the decision to publish the results.

References

1. Girhammar, U.A.; Pajari, M. Tests and Analysis on Shear Strength of Composite Slabs of Hollow Core Units and Concrete Topping. *Constr. Build. Mater.* **2008**, *22*, 1708–1722. [CrossRef]
2. Bertagnoli, G.; Mancini, G. Failure Analysis of Hollow-Core Slabs Tested in Shear. *Struct. Concr.* **2009**, *10*, 139–152. [CrossRef]
3. Palmer, K.D.; Schultz, A.E. Factors affecting web-shear capacity of deep hollow-core units. *PCI J.* **2010**, *55*, 123–146. [CrossRef]
4. Walraven, J.C.; Mercx, W.P.M. The bearing capacity of prestressed hollow-core slabs. *Heron* **1983**, *28*, 1–46.
5. Yang, L. Ein Haufwerksthe Oretisches Model der Restfestigkeit Grschadigter Betone. Ph.D. Thesis, The Technical University of Braunschweig, Braunschweig, Germany, 1978. (In German).
6. Pajari, M. *Design of Prestressed Hollow Core Slabs*; Research Reports 657; Technical Research Centre of Finland (VTT): Espoo, Finland, 1989.
7. Pajari, M. *Resistance of Prestressed Hollow Core Slabs against Web Shear Failure*; Technical Research Centre of Finland (VTT): Espoo, Finland, 2005.
8. Becker, R.J.; Buettner, D.R. Shear tests of extruded hollow-core slabs. *PCI J.* **1985**, *30*, 40–54. [CrossRef]
9. Mones, R.M.; Breña, S.F. Hollow-core slabs with cast-in-place concrete toppings: A study of interfacial shear strength. *PCI J.* **2013**, *58*, 124–141. [CrossRef]
10. Pajari, M. Web shear failure in prestressed hollow core slabs. *J. Struct. Eng.* **2009**, *42*, 83–104.
11. Im, J.H.; Park, M.K.; Lee, D.H.; Seo, S.Y.; Kim, K.S. Evaluation of horizontal shear strength of prestressed hollow-core slabs with cast-in-place topping concrete. *J. Korea Concr. Inst.* **2014**, *26*, 741–749. (In Korean) [CrossRef]
12. Moon, J.H. *Structural Performance for the Extruded Hollow Core Slabs with Prestressed Concretes*; Hansung PCC & Hannam University Research Report; Hansung PCC & Hannam University: Daejeon, Republic of Korea, 2017. (In Korean)

13. Lee, Y.J.; Kim, H.K.; Kim, M.J.; Lee, J.S.; Kim, K.H. Evaluation of structural performance for prestressed concrete hollow core slabs. *J. Korea Concr. Inst.* **2019**, *31*, 505–512. (In Korean) [CrossRef]
14. Jang, H.J.; Ryu, J.H.; Kim, S.M.; Ju, Y.K.; Kim, S.D. An evaluation on the flexural capacity of void slabs with prestressed half PC. *J. Archit. Inst. Korea* **2012**, *28*, 11–18. (In Korean)
15. Lee, Y.J.; Kim, H.K.; Kim, M.J.; Kim, D.H.; Kim, K.H. Shear Performance for Prestressed Concrete Hollow Core Slabs. *Appl. Sci.* **2020**, *10*, 1636. [CrossRef]
16. Khalil, A.A.; El Shafiey, T.F.; Mahmoud, M.H.; Baraghith, A.T.; Etman, A.E. Shear Behavior of Innovated Composite Hollow Core Slabs. In Proceedings of the International Conference on Advanced in Structural and Geotechnical Engineering (ICASGE), Hurghada, Egypt, 25–28 March 2019; pp. 1–12.
17. Araujo CA, M.; Loriggio, D.D.; Da Camara, J.M.M.N. Anchorage Failure and Shear Design of Hollow-core Slabs. *Struct. Concr.* **2011**, *12*, 109–119. [CrossRef]
18. ACI Committee 318. *Building Code Requirements for Structural Concrete (ACI 318-19) and Commentary (ACI 318R-19)*; American Concrete Institute (ACI): Farmington Hills, MI, USA, 2019; 623p.
19. *KDS 14 20 22*; MOLIT Shear and Torsion Design Code for Concrete Structures. Ministry of Land, Infrastructure and Transport, MOLIT: Sejong, Republic of Korea, 2021. (In Korean)
20. Precast/Prestressed Concrete Institute. *PCI Manual for the Design of Hollow Core Slabs*; PCI Hollow Core Slab Producers Committee: Chicago, IL, USA, 2017.
21. Hawkins, N.M.; Ghosh, S.K. Shear strength of hollow-core slabs. *PCI J.* **2006**, *51*, 110–114.
22. *ISO 1920-3*; Testing of Concrete-Part 3: Making and Curing Test Specimens. International Organization for Standardization: Geneva, Switzerland, 2019.
23. *ISO 1920-4*; Testing of Concrete-Part 4: Strength of Hardened Concrete. International Organization for Standardization: Geneva, Switzerland, 2020.
24. Kim, D.H.; Kim, M.J.; Jo, M.S.; Kim, H.G.; Yoon, Y.J.; Kim, K.H. Evaluation of Flexural Behavior of Prestressed Concrete (PSC) Hollow-Core Slabs (HCSs). *Buildings* **2023**, *13*, 2869. [CrossRef]
25. Palmer, K.D.; Schultz, A.E. Experimental investigation of the web-shear strength of deep hollow-core units. *PCI J.* **2011**, *56*, 83–104. [CrossRef]
26. TNO Building and Constructions Research. *TNO Report: Standard Shear Tests on Prestressed Hollow Core Slabs According to EN 1168*; TNO Building and Constructions Research: Hague, The Netherlands, 2005.
27. Celal, M.S. Shear Behavior of Precast/Prestressed Hollow-Core Slabs. Master's Thesis, University of Manitoba, Winnipeg, MB, Canada, 2011.
28. Park, M.K.; Lee, D.H.; Han, S.J.; Kim, K.S. Web-shear capacity of thick precast prestressed hollow-core slab units produced by extrusion method. *Int. J. Concr. Struct. Mater.* **2019**, *13*, 19–32. [CrossRef]

buildings

Article

The Synergistic Effect of Water Reducer and Water-Repellent Admixture on the Properties of Cement-Based Material

Raja Al jarmouzi [1,2], Zhenping Sun [1,2,*], Haijing Yang [1,2,*] and Yanliang Ji [1,2,3]

1 Key Laboratory of Advanced Civil Engineering Materials of Ministry of Education, Tongji University, Shanghai 201804, China; 1890002@tongji.edu.cn (R.A.j.); ji@tu-berlin.de (Y.J.)
2 School of Materials Science and Engineering, Tongji University, Shanghai 201804, China
3 Department of Civil Engineering, Technische Universität Berlin, 13355 Berlin, Germany
* Correspondence: szhp@tongji.edu.cn (Z.S.); yanghaijings@tongji.edu.cn (H.Y.)

Abstract: Water reducer and water-repellent admixture are very important in improving the workability and durability of cement-based materials. However, the synergistic effect of the two types of admixtures has not been well investigated. In this study, polycarboxylate ether-based superplasticizer (PCE) and octyltriethoxysilane (OTS) were adopted as water reducer and water-repellent admixture, respectively. Their synergistic effect on the fluidity, compressive strength, and water absorption rate of cement-based materials was investigated. Particularly, the pore structure and hydration state of cement paste were analyzed using ^{1}H Low-Field Nuclear Magnetic Resonance (^{1}H LF NMR). The result showed that the fluidity of cement paste containing different dosages of PCE was reduced by 5–10 mm by incorporating 1% OTS, and the compressive strength at the early age of 3 d of mortar containing high PCE dosage of 0.25% decreased up to 15% by using 1% OTS. In contrast, the compressive strength of mortar containing 0.20% PCE was slightly enhanced by the addition of 1% OTS. ^{1}H LF NMR analysis revealed that the combination of PCE and OTS would increase the pore size and total pore volume of cement paste, and more bleeding water would be generated at high PCE dosage. The intensity-weighted T_2 values of the main peak ($\overline{T_2}$) implied that both PCE and OTS produced a retardation effect on cement hydration. However, the water absorption rate decreased by 46.6% despite the increase in pore size and total pore volume. The conflict phenomenon powerfully revealed that the internal hydrophobic treatment by OTS has been successfully achieved. Overall, the combination of 0.20% PCE and 1% OTS exerted a positive synergistic effect in improving the compressive strength and water-repelling ability of cement-based materials, which is meaningful for improving their durability and service life.

Keywords: water-repellent admixture; internal hydrophobic treatment; pore structure; water absorption rate; bleeding

Citation: Al jarmouzi, R.; Sun, Z.; Yang, H.; Ji, Y. The Synergistic Effect of Water Reducer and Water-Repellent Admixture on the Properties of Cement-Based Material. *Buildings* **2024**, *14*, 2734. https://doi.org/10.3390/buildings14092734

Academic Editor: Geo Paul

Received: 25 July 2024
Revised: 25 August 2024
Accepted: 29 August 2024
Published: 31 August 2024

1. Introduction

Cement-based materials (CBM) are the most widely used building materials all over the world, depending on their superior advantages of versatility, high strength, and relatively low cost. During recent decades, great efforts have been made to improve the micro–macro properties of CBM to meet the increasing requirements in construction projects [1–4]. Regarding the hydrophilic and mesoporous properties of CBM [5–8], special attention has been paid to increasing their durability and service life, which declined due to the penetration of water and aggressive ions (magnesium, chloride, sulfate, etc.) into their porous microstructures and led to steel bars corrosion, leaching of hydration product, and freeze–thaw action [5,9–13]. The most effective method has been established, which is using silane-based compounds as water-repellent admixtures. Since the alkoxy group can easily hydrolyze in the alkaline environment of CBM, chemical bonds are formed between the silicate of cement and silane-based compounds; thus, a hydrophobic layer on the surface of cement particles is produced, which helps prevent the ingression of water and aggressive ions [7,14–21]. Wang et al. used polydimethylsiloxane (PDMS) for the hydrophobic modification of mortar. They found that the water absorption reduced approximately 92.51%, and the contact angle increased to 157.3° of the modified mortar [17]. Raja et al. studied the effect of octyltriethoxysilane, isobutyltriethoxysilane (IS), and waterborne pure nano-silicone emulsion (NS) as water-repellent admixtures on the water absorption and mechanical properties. The results exhibited that water absorption significantly reduced in the modified samples [22].

There are two main methods to apply silane-based water-repellent admixtures. The first way is the direct surface hydrophobic treatment [6,16,23–26]. This method can be applied to both new and old CBM surfaces. However, the hydrophobic property of CBM will be lost due to cracks, coating aging, and peels [9,14,17]. The second way is the internal hydrophobic treatment by incorporating silane-based water-repellent admixtures during the mixing procedure of fresh CBM to endow an integral hydrophobic property [9,14,17]. In general, the advantage of the internal hydrophobic treatment is that CBM is hydrophobic even with exposure to deterioration factors such as wear or cracking during service life [14,17–19].

Silane-based water-repellent admixtures can be used in the forms of dry powder [27,28] and liquid [16,17]. The most commonly used silane-based water-repellent admixtures include octyltriethoxysilane (OTS) and isobutyltriethoxysilane [20,29,30]. OTS is unique in the hydrophobization of CBM depending on its longer alkyl chain and excellent stability, which leads to destroying the original spatial correlation and weakening the capillary adsorption [31]. Furthermore, at room temperature, OTS is in liquid form that can be utilized directly or converted into an environmentally friendly water-based emulsion [6]. Studies showed that the use of OTS can increase the contact angle, reduce capillary water absorption, and improve the durability of CBM [20,32–34]. Xue et al. synthesized a water-based hydrophobic agent with OTS and then impregnated the surfaces of mortar and concrete. The result showed that the capillary water absorption of treated samples decreased by 94.6% compared to the un-modified one [30]. However, K. Grabowska et al. [32] reported that the compressive strength of mortar decreased by 15% using OTS. Y.G. Zhu et al. [34] used OTS for the internal hydrophobic treatment of recycled aggregate concrete. They reported that the compressive strength of concrete was 38% lower than that of un-modified samples. Specifically, the porosity increased at high OTS dosages [32,34,35].

Using a water reducer can be regarded as an effective way to increase the mechanical strength of CBM by effectively reducing the water-to-cement ratio. Polycarboxylate-based ester/ether superplasticizer (PCE) is widely used in CBM as a water reducer, which not only lowers water consumption but also improves workability, reduces cement consumption, increases mechanical strength, and enhances durability [36,37]. Several studies reported that PCE modified by alkoxysilane exhibited good dispersion, flowability retention, and low sulfate sensitivity in CBM [38–40]. C.A. Casagrande et al. [40] studied the effect of partial substitution of PCE by silane-based water-repellent admixtures in cement paste. They reported that using tetraethoxysilane (TEOS) for substitution could accelerate cement hydration as well as improve its workability and mechanical strength. However, aminoethylaminopropyl-trimethoxysilane (AEAPTMS) and 3-glycidoxypropyltrimethoxysilane (3-GPTMS) exhibited opposite trends. To the best of our knowledge, the synergistic effect of water reducer and water-repellent admixture on CBM has not been well studied so far.

In this work, a self-synthesized PCE was adopted as a water reducer, and OTS was utilized as a water-repellent admixture to achieve internal hydrophobicity. The synergistic effect of PCE and OTS on CBM was studied by evaluating the fluidity of cement paste, the compressive strength, and the water absorption rate of mortar. Moreover, the ^{1}H Low-Field Nuclear Magnetic Resonance (^{1}H LF NMR) was employed to investigate the pore structure and hydration state of cement paste. Through this research, a comprehensive understanding of the synergistic effect of water reducer and water-repellent admixture on

CBM can be drawn, which is meaningful to utilizing water-repellent admixture as well as improving the durability of CBM.

2. Materials and Methods

2.1. Materials

To reduce the interference of paramagnetic materials on the ^{1}H LF NMR test, a white Portland cement (P·W 42.5) with a low content (0.35%) of ferric oxide was used in this study. The oxide composition of cement was determined by an X-ray fluorescence spectrometer (PANalytical Axios), and the result is shown in Table 1. Self-synthesized PCE was used as a water reducer and characterized with gel permeation chromatography (GPC) by a Waters Alliance 2695 instrument (Waters, Eschborn, Germany), and the result is shown in Table 2. Octyltriethoxysilane (OTS) in chemical pure grade purchased from Macklin Co., Ltd. (Shanghai, China) was used as a water-repellent admixture. The structures of PCE and OTS are shown in Figure 1. The standard sand was supplied by Xiamen ISO Standard Sand Co., Ltd., Xiamen, China. The tap water was used to prepare cement paste and mortar.

Table 1. The oxide composition of P·W 42.5 used in this study (%).

CaO	SiO_2	MgO	SO_3	Al_2O_3	K_2O	Fe_2O_3	Na_2O	P_2O_5	Others
60.1	18.2	13.5	2.5	2.2	0.5	0.3	0.3	0.1	2.3

Table 2. Molecular masses (M_w and M_n), polydispersity index (PDI), and conversion rate of the macromonomer for the self-synthesized PCE.

M_w (Da)	M_n (Da)	PDI (M_w/M_n)	Conversion Rate of the Macromonomer (%)
58,910	24,980	2.4	90.0

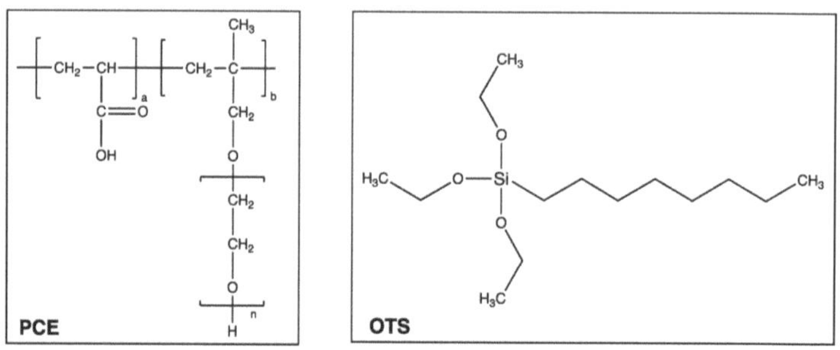

Figure 1. The molecular structure of PCE (a:b = 4.5, n = 23) and OTS.

2.2. Sample Preparation

In this study, fluidity test and pore structure determination were applied with cement paste, while compressive strength and water absorption rate were measured with mortar specimens. To prepare cement paste, 300 g of cement was used, and a low w/c of 0.25 was adopted. To prepare mortar, 450 g of cement and 1350 g of standard sand were used, and the w/c was 0.5. The dosages (mass percentage of PCE and OTS to cement) of PCE were 0.15%, 0.20%, and 0.25%, while the dosage of OTS was 1%. Throughout this paper, the following nomenclature was used to identify specimens: PCE_X-OTS_Y; X = 0.15, 0.20, and 0.25; Y = 0 and 1, representing the dosage of PCE and OTS, respectively. The mixing of cement paste and mortar was performed in accordance with Chinese standards GB/T1346-2001 [41] and GB/T 17671-2021 [42]. After the mixing process, the mortar specimens were cast into (40 × 40 × 40) mm^3 cubes and covered with a polytetrafluoroethylene film on the surface. For the first 24 h, the specimens were cured under the temperature of (20 ± 3) °C and the relative humidity of 90%. After that, the specimens were demolded and then cured continuously in the standard curing room until a certain age.

2.3. Fluidity Measurement

The fluidity of cement paste was determined using a mini-slump cone, with a height of 60 mm, an upper diameter of 36 mm, and a bottom diameter of 60 mm. For each sample, the fluidity result was taken from the average of three specimens, and the standard deviation was limited to 2%.

2.4. Pore Structure Determination

In this study, the pore structure of cement paste was determined using ^{1}H LF NMR apparatus (PQ-001, Niumag, Shanghai, China), with a 0.5 T magnetic field and a 25 mm radiofrequency coil. The T_2 transverse relaxation time was measured by the Carr–Purcell–Meiboom–Gill (CPMG) sequence. The number of scans and the echo time were kept constant at 4 and 0.302 ms, respectively. After mixing, fresh cement paste was cast into a small glass bottle and then placed in the apparatus; T_2 and relative amplitude were monitored at the age of 0, 1, 2, 3, 4, 5, 6, and 12 h and 1, 2, 3, 7, and 28 d. T_2 data were fitted to a multi-exponential curve using the Multi ExpInv Analysis software 4.0 (Niumag Electric Corporation, Shanghai, China), which was performed via the inverse Laplace transform algorithm [43]. The intensity-weighted T_2 values of the main peak ($\overline{T_2}$) calculated according to Equation (1) are used to characterize the hydration state, where I_i is the intensity at each T_2; t is the start point at T_2 distinguishing surface water peak and other peaks in T_2 distribution curves. For each sample, the result was taken from one specimen and the accuracy was examined by an extra specimen.

$$\overline{T_2} = \frac{\sum_{i=1}^{t} T_{2i} * I_i}{\sum_{i=1}^{t} I_i} \tag{1}$$

2.5. Compressive Strength Test

The compressive strength of mortar specimens at the ages of 3, 7, and 28 d were measured. The compression test machine with a capacity of 2000 kN was used at a loading rate of (2400 ± 200) N/s. For each sample, 3 specimens were measured, and their average value was calculated as the final result. The standard deviation is limited to 15%.

2.6. Water Absorption Rate Measurement

The water absorption rate was measured using the immersion method at the age of 7 d, considering that cement hydration was almost stopped [14,16,44]. The mortar specimen was dried at 50 °C for approximately three days to the constant weight M_a. After drying, the whole specimen was immersed in water, and the bottom of the specimen was separated from the container with a 3 mm thick plastic mesh to allow for water absorption. The water level was kept at 2–5 mm above the top of the specimen throughout the immersion time. The mass of the specimen M_C was recorded after immersed for 30 min and 1, 2, 3, 4, 5, 6, 12, 24, 48, and 72 h. Thus, the water absorption rate W_A was calculated according to Equation (2) [14,16]. Each result was the average value calculated with 3 specimens, and the standard deviation was limited to 2%.

$$W_A = \frac{M_C - M_a}{M_a} \times 100\% \tag{2}$$

3. Results

3.1. Fluidity

First, PCE dosages of 0.15%, 0.20%, and 0.25% were used to prepare cement paste, respectively. It can be seen from Figure 2 that the fluidity of cement paste increased significantly with increasing PCE. With the addition of 1% OTS, the fluidity of the cement paste decreased depending on PCE dosage. The reduction amount was about 10 mm when

PCE dosages were 0.15% and 0.25%, and 5 mm while PCE dosage was 0.20%. The result revealed that introducing OTS could impair the dispersion ability of PCE, thus reducing the fluidity of cement paste. This result is in agreement with the previous study [45]. The possible reason can be attributed to the fact that OTS hinders the adsorption of PCE on the surface of cement particles and hydration products.

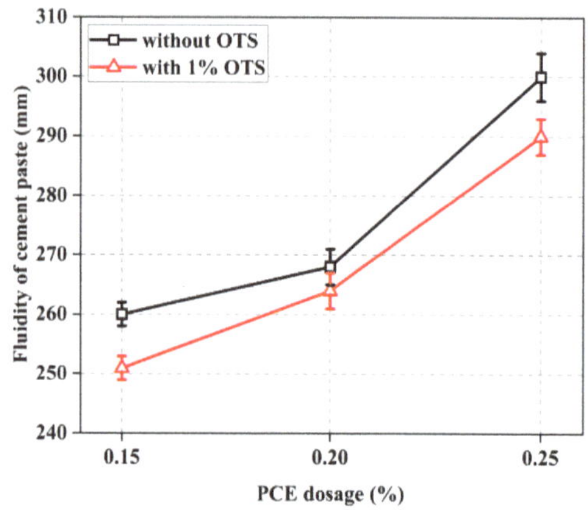

Figure 2. The fluidity of cement paste containing different dosages of PCE without and with the presence of 1% OTS.

3.2. T_2 Relaxation Time Distributions

The transverse relaxation time (T_2) correlates linearly with the volume-to-surface ratio of pores in terms of pore diameter. In addition, the cumulative area under the T_2 curve can quantitatively represent the relative pore volume in the cement paste. Hence, the T_2 distribution reflects the pore structure of cement paste [43,46]. Previous research has revealed that short T_2 values in the range of 0.01–1 ms represent the water in the gel pores of about 3 nm, conceptualized as existing between the gel globules that comprise C-S-H. Longer T_2 values in the range of 1–100 ms correspond to the water in the capillary pores, which initially is interstitial water between the clinker grains but then evolves to become inter-hydrate water. Further, T_2 values in the range of 100–10,000 ms can be attributed to water in the meso and macro pores, which can be regarded as bleeding water [43,47,48].

Figure 3 exhibits the T_2 distributions of cement paste mixed with 0.15%, 0.20%, and 0.25% PCE without and with 1% OTS. It can be seen that the relative amplitude of the T_2 main peaks decreased and shifted to the left overall as a function of time, indicating that the pore volume of cement paste reduced and the pore diameter became smaller. Special attention must be paid to the weak peak appearing around 1000 ms. By comparing Figure 3a,c,e, it can be found that the weak peak around 1000 ms arose earlier, and the intensity increased with increasing PCE dosage, revealing that the bleeding water came out earlier and its amount increased with more PCE added. These weak peaks almost faded before 12 h due to the reabsorption of water by cement paste and its involvement in the hydration reaction. From Figure 3b,d,f, it can be observed that with the addition of 1% OTS, the weak peak around 1000 ms appeared immediately after the mixing procedure and could be captured even after 1 d. Furthermore, the intensity of the weak peak around 1000 ms increased significantly compared to that without the addition of 1% OTS. The result disclosed that with the addition of 1% OTS, a strong water-repelling property was created in cement paste due to the hydrophobic effect of OTS. Therefore, the bleeding water appeared from the very beginning, and its migration from the surface to the interior part was hindered as well. As more bleeding water was presented, the water within the cement paste for dispersion was reduced; thus, the fluidity was decreased with the addition of 1% OTS. Notably, the water-to-cement ratio would be increased on the surface of cement paste, and the large-scale pores would be generated because of the increased bleeding water [43,49], which would further affect hydration product distribution, microstructure development, and mechanical properties [50–52].

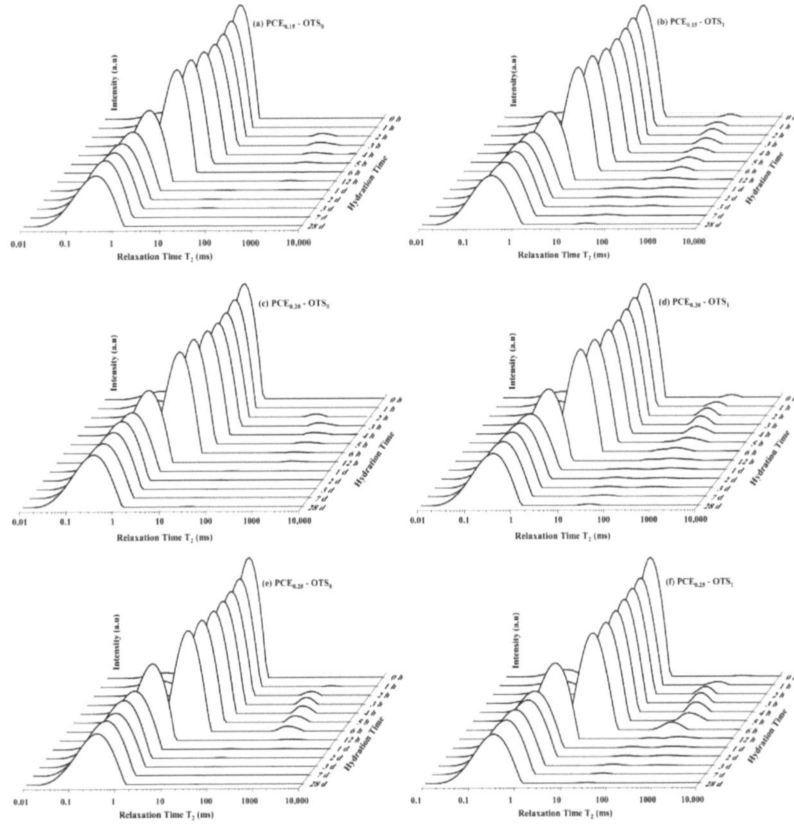

Figure 3. The T_2 distribution of cement paste: (**a**) $PCE_{0.15}$-OTS_0; (**b**) $PCE_{0.15}$-OTS_1; (**c**) $PCE_{0.20}$-OTS_0; (**d**) $PCE_{0.20}$-OTS_1; (**e**) $PCE_{0.25}$-OTS_0; and (**f**) $PCE_{0.25}$-OTS_1.

H. Liu et al. [53] have studied the pore structure development and water migration process of cement paste using ^{1}H LF NMR by analyzing intensity-weighted T_2 values ($\overline{T_2}$), which were calculated according to Equation (1). Based on their research, the decrease in $\overline{T_2}$ indicates that the average pore diameter of cement paste reduced, which can be attributed to that the hydration product has filled spaces between particles. Further, as more water participates in hydration, free water migrates from macropores to micropores, causing $\overline{T_2}$ decrease as well. To better illustrate the pore structure development and hydration state of cement paste, the $\overline{T_2}$ was calculated further in the same manner as described by H. Liu et al. [53], and the results are shown in Figure 4. As the T_2 distribution curves do not change further after 1 d in Figure 3, the $\overline{T_2}$ were collected in the time range of 0–24 h, correspondingly. Figure 4 shows the $\overline{T_2}$ of cement paste mixed with 0.15%, 0.20%, and 0.25% PCE, respectively, without and with 1% OTS as a function of hydration time. It can be observed that the $\overline{T_2}$ decreased at different rates as hydration time prolonged. Specifically, the $\overline{T_2}$ increased with increasing dosage of PCE at 0–3 h, indicating that the pore diameter was increased. As pores were filled with free water, the result implies that more free water was released by adding more PCE, and less water participated in cement hydration, indicating that the cement hydration was delayed with the addition of more PCE. During (3–12) h, the $\overline{T_2}$ reduced sharply; the more the PCE was added, the higher the reduction rate was, indicating the faster hydration rate. Furthermore, with the addition of 1% OTS, the $\overline{T_2}$ increased obviously, signifying that the pore diameter increased. Therefore, it can be inferred that OTS increased the amount of free water and may cause a delay in cement hydration. After 12 h, the $\overline{T_2}$ almost reached a plateau, indicating that the variation in pore diameter and the migration in free water slowed down.

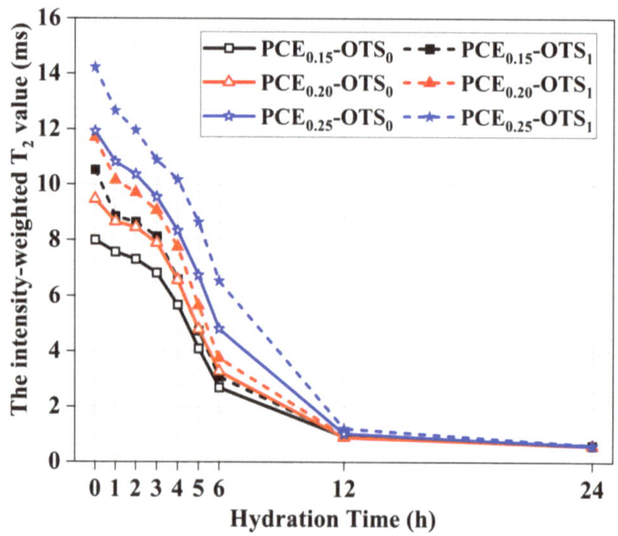

Figure 4. The $\overline{T_2}$ of cement paste containing different dosages of PCE without and with the presence of 1% OTS.

In general, the cumulative peak area of the T_2 spectrum can relatively represent the amount of free water existing in the pores of cement paste, which further reflects the total pore volume. Figure 5 exhibits the cumulative peak area of the T_2 spectrum of cement paste mixed with 0.15%, 0.20%, and 0.25% PCE, respectively, without and with 1% OTS. It can be seen that the cumulative peak area decreased continuously within (0–24) h, indicating that the amount of free water reduced gradually due to the hydration reaction, in which free water was converted into chemically bonded water, and the total porosity reduced as well. Furthermore, the cumulative peak area increased slightly with increasing PCE dosage as well as the addition of 1% OTS. The possible reason behind this result can be attributed to two aspects. The first reason was that more free water was released due to the enhanced dispersing ability of increased PCE and hydrophobicity of OTS. Another reason was that both PCE and OTS provided a retarding effect to cement hydration; thus, the water consumption was reduced.

Overall, the following ideas can be drawn from the ^{1}H LF NMR study:

- The bleeding water can be generated due to the addition of PCE and OTS. The higher the dosages of PCE, the more the bleeding water;
- The calculated $\overline{T_2}$ revealed that the addition of PCE and OTS provided a retarding effect upon cement hydration. And this retarding effect weakened gradually with prolonged hydration time;
- The porosity of cement paste would be increased with the addition of PCE and OTS.

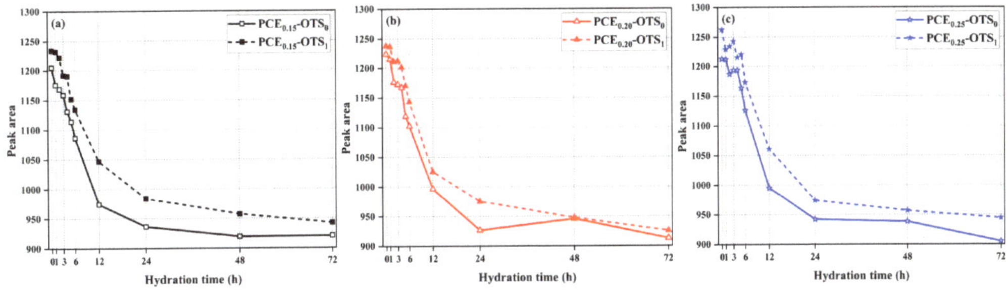

Figure 5. The cumulative peak area of T_2 spectrum of cement paste containing different dosages of PCE without and with the presence of 1% OTS: (**a**) 0.15% PCE; (**b**) 0.20% PCE; (**c**) 0.25% PCE.

3.3. Compressive Strength

The compressive strength of mortar specimens containing 0.15%, 0.20%, and 0.25% PCE were measured without and with the presence of 1% OTS. The result is summarized in Figure 6. By incorporating 1% OTS, the compressive strength decreased at PCE dosages of 0.15% and 0.25%. At a PCE dosage of 0.15%, the compressive strength decreased by 6.2%, 5.5%, and 2.2% at the age of 3 d, 7 d, and 28 d with the addition of 1% OTS, respectively. At low dosages, the dispersion ability of PCE was easily weakened by introducing OTS, thus reducing the workability of mortar and causing defections during the casting procedure, which were harmful to compressive strength. Higher reduction rates in compressive strength by adding 1% OTS were found at the early ages of 3 d and 7 d, while PCE dosage increased up to 0.25%, which were 15.0% and 9.7%, respectively. Inferred from the ^{1}H LF NMR result, this phenomenon could be attributed to bleeding, which happened in the first 6 h. The large-scale pore size on the surface and the movement of cement particles during the bleeding process led to uneven distribution and more defects, which resulted in a significant decrease in compressive strength. This result is in agreement with previous studies [50,51]. In contrast, an enhancing effect of 1% OTS was observed while PCE dosage was 0.20%, indicating that 0.20% PCE and 1% OTS were an adequate combination, which made mortar obtain a better workability and compact structure, as is revealed in Sections 3.1 and 3.2. Finally, it can be concluded that the synergistic of PCEs and OTS at adequate contents can remarkably increase the compressive strength of mortar, which allows for more applications of water-repellent-based saline in hydrophobic concrete in different areas exposed to water actions.

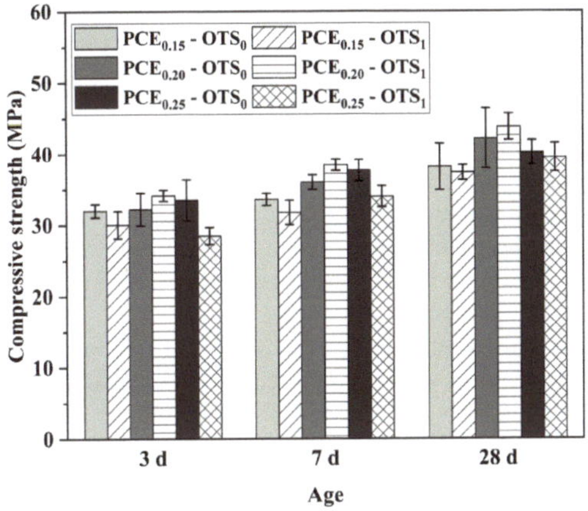

Figure 6. The compressive strength of mortar containing different dosages of PCE without and with the presence of 1% OTS.

3.4. Water Absorption Rate

Figure 7 presents the water absorption rate of mortar containing different dosages of PCE without and with the presence of 1% OTS. It can be observed that the water absorption rate increased continuously with the prolonged immersion duration. When the PCE dosage was increased from 0.15% to 0.20%, the water absorption rate was obviously reduced However, the water absorption rate increased drastically when PCE dosage increased to 0.25%. The high content of PCEs in mortar leads to a bleeding effect, as is shown in the ^{1}H LF NMR result. During the bleeding process, the top layer of hydrated mortar is more porous than the lower layer due to bleeding, which then increases the water absorption rate of hardened cement mortar [50,51].

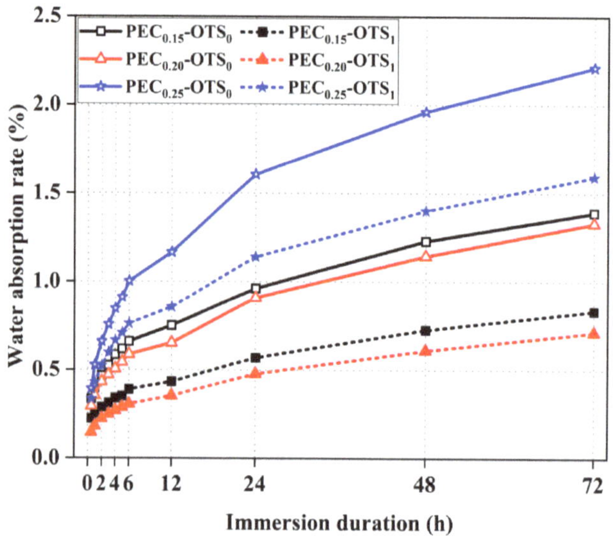

Figure 7. The water absorption rate of mortar containing different dosages of PCE without and with the presence of 1% OTS.

Cheerfully, the addition of 1% OTS provided a significant effect in reducing the water absorption rate. It can be observed in Figure 8 that after being immersed for 72 h, with the addition of 1% OTS, the water absorption rate decreased by 40.3%, 46.6%, and 28.1% in the mortar containing 0.15%, 0.20%, and 0.25% PCE, respectively. As is shown in the ^{1}H LF NMR result, the addition of OTS increased the pore size and total pore volume of cement paste; it seems that an increased water absorption rate would be produced under these effects. However, the reverse result showed that the water absorption rate reduced significantly with the addition of OTS. The conflict phenomenon powerfully revealed that the internal hydrophobic treatment by OTS has been successfully achieved. When OTS was incorporated into mortar during the mixing procedure, the interaction between OTS molecule and cement, as well as hydration products, led to the formation of a stable self-assembled hydrophobic film on the surface of the pores and on the external surface of the mortar, which provided a high water-repellent ability [7,10]. The lowest water absorption rate was obtained when 0.20% PCE and 1% OTS were used, indicating the most excellent synergistic effect in water repelling. It can be predicted that the synergistic effect of PCE and OTS can remarkably improve the durability of cement-based materials and allow for more applications of water-repellent-based saline in cement-based materials.

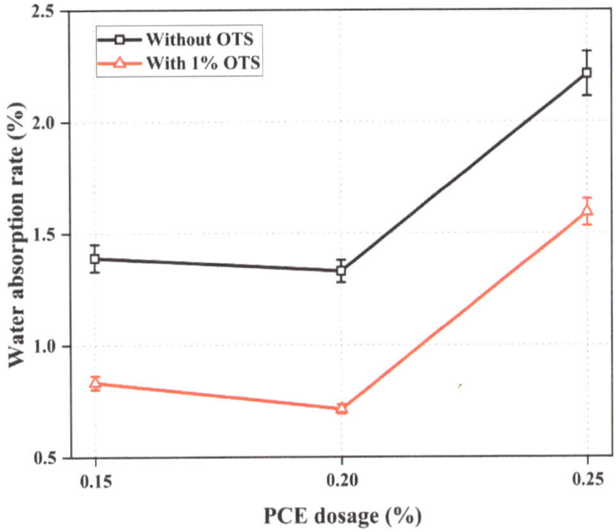

Figure 8. The water absorption rate of mortar containing different dosages of PCE without and with the presence of 1% OTS after being immersed in water for 72 h.

4. Conclusions

Water reducer and water-repellent admixture are two important components for improving the workability and durability of CBM, respectively. However, when the two species of admixtures are incorporated at the same time, their synergistic effect may generate a profound influence on the basic properties of CBM, which have not been studied extensively. In this study, by incorporating PCE and OTS in the mixing procedure to prepare CBM, the fluidity, pore structure, compressive strength, and water absorption rate of CBM were investigated. The following main conclusions can be drawn:

- The fluidity of cement paste was reduced with the addition of OTS for about 5–10 mm, especially at a low PCE dosage. On the one hand, OTS impairs the dispersion ability of PCE as OTS may hinder the adsorption of PCE on the surface of cement particles and a hydration product. On the other hand, OTS caused more bleeding water, thus reducing the free-water amount within the cement paste for dispersion;
- The ^{1}H LF NMR results showed that incorporating PCE and OTS resulted in an increased pore size as well as total pore volume. In particular, more bleeding water was presented by using OTS. The intensity-weighted T_2 values ($\overline{T_2}$) revealed that the addition of PCE and OTS provided a retardation effect on cement hydration;
- Mortar compressive strength decreased by 15.0% at the age of 3 d with the addition of OTS at a high PCE dosage of 0.25%, which could be attributed to a more porous top layer resulting from increased bleeding water amount. Notably, a positive synergistic effect of PCE and OTS was endowed on the compressive strength when PCE dosage was 0.20%;
- Despite the increased pore size and pore volume, as well as the retarded cement hydration, the water absorption rate was still reduced up to 46.6% with the addition of OTS. The conflict phenomenon powerfully revealed that the internal hydrophobic treatment by OTS has been successfully achieved.

The results obtained from this research clearly reveal the synergistic effect of water reducer and water-repellent admixture on the basic properties of CBM and enrich our knowledge in using water-repellent admixture through internal treatment, which is meaningful for improving the durability of cement-based materials.

Author Contributions: Conceptualization, R.A.j., Z.S., H.Y. and Y.J.; methodology, R.A.j., Z.S., H.Y. and Y.J.; software, R.A.j. and Y.J.; validation, R.A.j. and H.Y.; formal analysis, R.A.j. and H.Y.; investigation, R.A.j. and H.Y.; resources, Z.S., H.Y. and Y.J.; data curation, R.A.j. and H.Y.; writing—original draft preparation, R.A.j.; writing—review and editing, H.Y.; visualization, R.A.j. and H.Y.; supervision, Z.S.; project administration, Z.S.; funding acquisition, Z.S. and Y.J. All authors have read and agreed to the published version of the manuscript.

Funding: This research was funded by the National Natural Science Foundation of China (Grant No. 52278272 and 52108240), the Science and Technology Commission of Shanghai Municipality (Grant No. 23DZ1203500), the Housing and Urban–Rural Construction Management Commission of Shanghai Municipality (Grant. No. 2021-001-002), the Science and Technology Plan Project of Inner Mongolia Autonomous Region (Grant. No. 2022YFDZ0063), the Expert Workstation Project of Science and Technology Department of Yunnan Province (Grant. No. 202105AF150243).

Data Availability Statement: The original contributions presented in the study are included in the article, further inquiries can be directed to the corresponding author.

Conflicts of Interest: The authors declare no conflicts of interest.

References

1. Kawashima, S.; Hou, P.; Corr, D.J.; Shah, S.P. Modification of cement-based materials with nanoparticles. *Cem. Concr. Compos.* **2013**, *36*, 8–15. [CrossRef]
2. Wu, H.; Gao, J.; Liu, C.; Zhao, Y.; Li, S. Development of nano-silica modification to enhance the micro-macro properties of cement-based materials with recycled clay brick powder. *J. Build. Eng.* **2024**, *86*, 108854. [CrossRef]
3. Musale, A.; Hunashyal, A.M.; Patil, A.Y.; Kumar, R.; Ahamad, T.; Kalam, M.A.; Patel, M. Study on nanomaterials coated natural coir fibers as crack arrestor in cement composite. *Adv. Civ. Eng.* **2024**, *2024*, 6686655. [CrossRef]
4. Roopa, A.K.; Hunashyal, A.M.; Patil, A.Y.; Kamadollishettar, A.; Patil, B.; Soudagar, M.E.M.; Shahapurkar, K.; Khan, T.M.Y.; Kalam, M.A. Study on interfacial interaction of cement-based nanocomposite by molecular dynamic analysis and an RVE approach. *Adv. Civ. Eng.* **2023**, *2023*, 8404335. [CrossRef]
5. Hewlett, P.C.; Liska, M. *Lea's Chemistry of Cement and Concrete*, 5th ed.; Butterworth-Heinemann: Cambridge, UK, 2019.
6. Song, J.L.; Li, Y.X.; Xu, W.; Liu, H.; Lu, Y. Inexpensive and non-fluorinated superhydrophobic concrete coating for anti-icing and anti-corrosion. *J. Colloid Interface Sci.* **2019**, *541*, 86–92. [CrossRef] [PubMed]
7. Ma, Z.M.; Zhu, F.Z.; Zhao, T.J. Effects of surface modification of silane coupling agent on the properties of concrete with freeze-thaw damage. *KSCE J. Civ. Eng.* **2018**, *22*, 657–669. [CrossRef]
8. Liu, G.; Zhao, T.; Fei, H.; Li, F.; Guo, W.; Yao, Z.; Feng, Z. A review of various self-cleaning surfaces, durability and functional applications on building exteriors. *Constr. Build. Mater.* **2023**, *409*, 134084. [CrossRef]
9. Feng, Z.J.; Wang, F.J.; Xie, T.; Ou, J.F.; Xue, M.S.; Li, W. Integral hydrophobic concrete without using silane. *Constr. Build. Mater.* **2019**, *227*, 116678. [CrossRef]
10. Liu, S.; Yang, J.; Chen, X.H.; Yang, G.T.; Cai, D.G. Application of mastic asphalt waterproofing layer in high-speed railway track in cold regions. *Appl. Sci.-Basel* **2018**, *8*, 667. [CrossRef]
11. Li, W.T.; Pour-Ghaz, M.; Castro, J.; Weiss, J. Water absorption and critical degree of saturation relating to freeze-thaw damage in concrete pavement joints. *J. Mater. Civ. Eng.* **2012**, *24*, 299–307. [CrossRef]
12. Yuan, Q.; Xie, Z.; Tian, Y.; Sun, L.M.; Yang, C.; Chi, H.; Zhu, X. The driving force of water absorption in cementitious materials: An analysis of surface-free energies and pore structure. *J. Mater. Res. Technol.* **2024**, *30*, 4714–4724. [CrossRef]
13. Pang, Y.; Wang, H.; Yang, L.; Tang, Q.; Li, H.; Zhang, J. Experimental study on freeze-thaw resistance of mortar: An attempt to modify hydrophobic materials with hydrophobic nano-silica. *J. Build. Eng.* **2024**, *95*, 110152. [CrossRef]
14. Liu, P.; Chen, Y.; Wang, W.L.; Yu, Z.W. Effect of physical and chemical sulfate attack on performance degradation of concrete under different conditions. *Chem. Phys. Lett.* **2020**, *745*, 137254. [CrossRef]
15. Bader, T.; Waldner, B.J.; Unterberger, S.H.; Lackner, R. On the performance of film formers versus penetrants as water-repellent treatment of (HPC) surfaces. *Constr. Build. Mater.* **2019**, *203*, 481–490. [CrossRef]
16. Subbiah, K.; Park, D.J.; Lee, Y.S.; Velu, S.; Lee, H.S.; Jang, H.O.; Choi, H.J. Development of water-repellent cement mortar using silane enriched with nanomaterials. *Prog. Org. Coat.* **2018**, *125*, 48–60. [CrossRef]
17. Wang, F.J.; Lei, S.; Ou, J.F.; Li, W. Effect of PDMS on the waterproofing performance and corrosion resistance of cement mortar. *Appl. Surf. Sci.* **2020**, *507*, 145016. [CrossRef]
18. Falchi, L.; Zendri, E.; Müller, U.; Fontana, P. The influence of water-repellent admixtures on the behaviour and the effectiveness of Portland limestone cement mortars. *Cem. Concr. Compos.* **2015**, *59*, 107–118. [CrossRef]
19. Li, F.P.; Liu, L.S.; Liu, K.; Zheng, A.H.; Liu, J.S. Investigation on waterproof mechanism and micro-structure of cement mortar incorporated with silicane. *Constr. Build. Mater.* **2020**, *239*, 117865. [CrossRef]

20. Grabowska, K.B.; Koniorczyk, M. Internal hydrophobization of cementitious materials by using of organosilicon compounds. In Proceedings of the 12th Nordic Symposium on Building Physics (Nsb 2020), Tallinn, Estonia, 6–9 September 2020; Volume 172, p. 14006. [CrossRef]
21. Tang, D.; Yang, C.; Shen, C.; Li, Z.; Lyu, Z.; Yu, L.; Huang, Q.; Zhu, X. Inhibiting efflorescence of alkali-activated slag mortar by improving its hydrophobicity with methyl-terminated polydimethylsiloxane (PDMS): Evidences showing the remained PDMS. *Constr. Build. Mater.* **2024**, *443*, 137688. [CrossRef]
22. Raja, A.; Sun, Z.P.; Ji, Y.L.; Yang, J.B. The influence of water-repellent admixtures on the hydration process, water absorption, and mechanical properties of cement paste: An NMR study combined with additional methods. *J. Mater. Civ. Eng.* **2022**, *34*, 04022267. [CrossRef]
23. Safiuddin, M. Concrete damage in field conditions and protective sealer and coating systems. *Coatings* **2017**, *7*, 90. [CrossRef]
24. Wang, F.J.; Xie, T.; Ou, J.F.; Xue, M.S.; Li, W. Cement based superhydrophobic coating with excellent robustness and solar reflective ability. *J. Alloys Compd.* **2020**, *823*, 153702. [CrossRef]
25. Li, Y.; Gou, L.; Wang, H.Q.; Wang, Y.Z.; Zhang, J.F.; Li, N.; Hu, S.L.; Yang, J.L. Fluorine-free superhydrophobic carbon-based coatings on the concrete. *Mater. Lett.* **2019**, *244*, 31–34. [CrossRef]
26. Kong, X.Q.; Shen, Y.D.; Shi, J.R.; Zhang, N.; Kang, R.; Fu, Y. Superhydrophobic concrete coating with excellent mechanical robustness and anti-corrosion performances. *Colloid Surface A* **2024**, *684*, 133157. [CrossRef]
27. Qu, Z.Y.; Yu, Q.L. Synthesizing super-hydrophobic ground granulated blast furnace slag to enhance the transport property of lightweight aggregate concrete. *Constr. Build. Mater.* **2018**, *191*, 176–186. [CrossRef]
28. Wong, H.S.; Barakat, R.; Alhilali, A.; Saleh, M.; Cheeseman, C.R. Hydrophobic concrete using waste paper sludge ash. *Cem. Concr. Res.* **2015**, *70*, 9–20. [CrossRef]
29. Yao, H.; Xie, Z.L.; Huang, C.H.; Yuan, Q.; Yu, Z.W. Recent progress of hydrophobic cement-based materials: Preparation, characterization and properties. *Constr. Build. Mater.* **2021**, *299*, 124255. [CrossRef]
30. Xue, X.; Li, Y.W.; Yang, Z.; He, Z.Y.; Dai, J.G.; Xu, L.J.; Zhang, W.D. A systematic investigation of the waterproofing performance and chloride resistance of a self-developed waterborne silane-based hydrophobic agent for mortar and concrete. *Constr. Build. Mater.* **2017**, *155*, 939–946. [CrossRef]
31. Zhang, W.J.; Li, S.C.; Hou, D.S.; Geng, Y.J.; Zhang, S.L.; Yin, B.; Li, X.G. Study on unsaturated transport of cement-based silane sol coating materials. *Coatings* **2019**, *9*, 427. [CrossRef]
32. Grabowska, K.; Koniorczyk, M. Influence of organosilicon admixtures on the hydration of Portland cement. *J. Therm. Anal. Calorim.* **2022**, *147*, 6131–6145. [CrossRef]
33. Vdovin, E.; Mavliev, L.; Stroganov, V. Interaction of clay soil components with Portland cement and complex additive based on octyltriethoxysilane and sodium hydroxide. *IOP Conf. Ser. Mater. Sci. Eng.* **2020**, *890*, 012031. [CrossRef]
34. Zhu, Y.G.; Kou, S.C.; Poon, C.S.; Dai, J.G.; Li, Q.Y. Influence of silane-based water repellent on the durability properties of recycled aggregate concrete. *Cem. Concr. Compos.* **2013**, *35*, 32–38. [CrossRef]
35. Bader, T.; Unterberger, S.H.; Lackner, R. Effect of substrate moisture on the weatherability of surface treatment for High-Performance Concrete (HPC). *Cem. Concr. Compos.* **2017**, *83*, 57–65. [CrossRef]
36. Huang, H.L.; Qian, C.X.; Zhao, F.; Qu, J.; Guo, J.Q.; Danzinger, M. Improvement on microstructure of concrete by polycarboxylate superplasticizer (PCE) and its influence on durability of concrete. *Constr. Build. Mater.* **2016**, *110*, 293–299. [CrossRef]
37. Ma, Y.H.; Bai, J.J.; Shi, C.J.; Sha, S.N.; Zhou, B.B. Effect of PCEs with different structures on hydration and properties of cementitious materials with low water-to-binder ratio. *Cem. Concr. Res.* **2021**, *142*, 106343. [CrossRef]
38. He, Y.; Zhang, X.; Hooton, R.D. Effects of organosilane-modified polycarboxylate superplasticizer on the fluidity and hydration properties of cement paste. *Constr. Build. Mater.* **2017**, *132*, 112–123. [CrossRef]
39. Fan, W.; Stoffelbach, F.; Rieger, J.; Regnaud, L.; Vichot, A.; Bresson, B.; Lequeux, N. A new class of organosilane-modified polycarboxylate superplasticizers with low sulfate sensitivity. *Cem. Concr. Res.* **2012**, *42*, 166–172. [CrossRef]
40. Casagrande, C.A.; Jochem, L.F.; Onghero, L.; de Matos, P.R.; Repette, W.L.; Gleize, P.J.P. Effect of partial substitution of superplasticizer by silanes in Portland cement pastes. *J. Build. Eng.* **2020**, *29*, 101226. [CrossRef]
41. *GB/T 1346-2001*; Test Methods fot Water Requirement of Normal Consistency, Setting Time and Soundness of the Portland Cements. General Administration of Quality Supervision, Inspection and Quarantine of the People's Republic of China Press: Beijing, China, 2011.
42. *GB/T 17671-2021*; Test Methods of Cement Mortar Strength (ISO Method). Standardization Administration of the People's Republic of China Press: Beijing, China, 2021.
43. Ji, Y.L.; Sun, Z.P.; Yang, J.B.; Pel, L.; Raja, A.; Ge, H.S. NMR study on bleeding properties of the fresh cement pastes mixed with polycarboxylate (PCE) superplasticizers. *Constr. Build. Mater.* **2020**, *240*, 117938. [CrossRef]
44. Elkhadiri, I.; Palacios, M.; Puertas, F. Effect of curing temperature on hydration process of different cement. *Ceramics-Silikaty* **2009**, *53*, 65–75.
45. Felekoglu, B. Rheological behaviour of self-compacting micro-concrete. *Sadhana* **2014**, *39*, 1471–1495. [CrossRef]
46. Fischer, N.; Haerdtl, R.; McDonald, P.J. Observation of the redistribution of nanoscale water filled porosity in cement based materials during wetting. *Cem. Concr. Res.* **2015**, *68*, 148–155. [CrossRef]
47. Muller, A.C.A.; Scrivener, K.L.; Gajewicz, A.M.; McDonald, P.J. Densification of C-S-H Measured by ^{1}H NMR Relaxometry. *J. Phys. Chem. C* **2013**, *117*, 403–412. [CrossRef]

48. Yuan, Q.; Zhong, F.W.; Zuo, S.H.; Xie, Z.L.; Xue, K.W.; Yao, H. Understanding the water transport behaviors of hydrophobic cement mortar by paraffin modification. *Constr. Build. Mater.* **2024**, *435*, 136845. [CrossRef]
49. Massoussi, N.; Keita, E.; Roussel, N. The heterogeneous nature of bleeding in cement pastes. *Cem. Concr. Res.* **2017**, *95*, 108–116. [CrossRef]
50. Li, H.; Mu, R.; Qing, L.B.; Chen, H.S.; Ma, Y.F. The influence of fiber orientation on bleeding of steel fiber reinforced cementitious composites. *Cem. Concr. Compos.* **2018**, *92*, 125–134. [CrossRef]
51. Feng, H.J.; Le, H.T.N.; Wang, S.S.; Zhang, M.H. Effects of silanes and silane derivatives on cement hydration and mechanical properties of mortars. *Constr. Build. Mater.* **2016**, *129*, 48–60. [CrossRef]
52. Guo, X.L.; Shi, H.S.; Xia, M. Geopolymerization process at early age by low field nuclear magnetic resonance. *J. Chin. Ceram. Soc.* **2015**, *43*, 138–143. [CrossRef]
53. Liu, H.; Sun, Z.P.; Yang, J.B.; Ji, Y.L. A novel method for semi-quantitative analysis of hydration degree of cement by ^{1}H low-field NMR. *Cem. Concr. Res.* **2021**, *141*, 106329. [CrossRef]

buildings

MDPI

Article

Microscopic Transport and Degradation Behavior of CO_2 in C-S-H with Varying Ca/Si Ratios during Carbonation

Wangzhe Sun [1], Chengbo Liu [1], Fen Hong [1], Pan Wang [1], Yue Zhang [1], Xinpeng Wang [1], Dongshuai Hou [1,2] and Muhan Wang [1,2,3,*]

1 Department of Civil Engineering, Qingdao University of Technology, Qingdao 266033, China; wangzhesun6@gmail.com (W.S.); liuchengbo0407@126.com (C.L.); hongf630220@163.com (F.H.); wangpan@qut.edu.cn (P.W.); zhangyue@qut.edu.cn (Y.Z.); wangxinpeng@qut.edu.cn (X.W.); dshou@outlook.com (D.H.)
2 Engineering Research Center of Concrete Technology under Marine Environment, Ministry of Education, Qingdao 266520, China
3 State Key Laboratory of Hydraulic Engineering Simulation and Safety, Tianjin University, Tianjin 300072, China
* Correspondence: wangmuhan@qut.edu.cn

Abstract: Carbonation is a critical factor contributing to the degradation of reinforced concrete systems. Understanding the micro-mechanism of concrete carbonation is essential for mitigating corrosion losses. This study investigates the transport and reaction processes of water and CO_2 in CSH pores with varying calcium–silica ratios using reactive force field molecular dynamics. Simulation results reveal that CO_2 and its hydration products occupy adsorption sites on the CSH, hindering solution transport within the pores. As the Ca/Si ratio increases, the adsorption of Ca ions on the CSH matrix weakens, facilitating Ca's reaction with CO_2 and its displacement from the CSH surface. Consequently, a wider distribution of Ca on the surface occurs, and CO_2 directly adsorbs onto the CSH matrix, widening the transport space and accelerating transport speed. Furthermore, the impact of bridging silica–oxygen on the CSH surface is analyzed, indicating that the absence of bridging silica–oxygen enhances adsorption sites for Ca ions, thus intensifying their adsorption on CSH.

Keywords: concrete durability; carbonation corrosion; pore transport; molecular dynamics; nanoscale properties

Citation: Sun, W.; Liu, C.; Hong, F.; Wang, P.; Zhang, Y.; Wang, X.; Hou, D.; Wang, M. Microscopic Transport and Degradation Behavior of CO_2 in C-S-H with Varying Ca/Si Ratios during Carbonation. *Buildings* **2024**, *14*, 2808. https://doi.org/10.3390/buildings14092808

Academic Editor: Hailong Ye

Received: 1 August 2024
Revised: 29 August 2024
Accepted: 3 September 2024
Published: 6 September 2024

1. Introduction

Carbonation significantly compromises the performance of concrete structures [1–5]. Primarily, it induces a decrease in the alkalinity of concrete, triggering depassivation of the reinforcement's protective film and subsequent corrosion [6–8]. This corrosion yields products two to six times larger in volume than the original iron, leading to concrete expansion, cracking, and exacerbating structural damage [9–12]. Additionally, carbonation contributes to matrix deterioration, as CO_2 must traverse within the matrix to reach reinforcement surfaces, where it further reacts, thereby diminishing concrete strength [2,3,13]. Given concrete's status as the most widely utilized engineered material, concerns regarding its durability are paramount. The substantial annual economic losses attributed to carbonation-induced degradation underscore the urgent need for comprehensive studies on carbon dioxide transport within concrete matrices [11,14].

As outlined, this study comprises two distinct facets: CO_2 transport and its subsequent reaction with the matrix. Firstly, concrete, being porous, facilitates CO_2 intrusion into its internal structure [13,15]. The rate and depth of this transport are influenced by factors such as CO_2 concentration, purity, gas pressure, environmental conditions (temperature, humidity), carbonation duration, and concrete composition [16–20]. Secondly, during transmission, CO_2 dissolves in the concrete pore solution, yielding $CO_3{}^{2-}$ and $HCO_3{}^-$, subsequently reacting with $Ca(OH)_2$ to form $CaCO_3$ and free water [1,2]. This carbonation reaction alters the microstructure within voids, reducing total concrete porosity and thereby potentially enhancing concrete strength, but at the expense of increased capillary porosity [21].

Furthermore, it lowers the concrete pH to approximately 9, leading to the dissolution of chemically bound chlorides in Friedel's salt, consequently releasing chlorides into the aqueous phase [22,23]. This combined effect renders the reinforcement more susceptible to corrosion. If exposed to a carbonated environment over an extended period, carbon dioxide will additionally react with CSH gel, depleting calcium hydroxide and forming amorphous silica gel [24]. This scenario is highly detrimental to reinforced concrete systems, underscoring the critical importance of studying carbonation transport and reaction processes to mitigate concrete carbonation. Mi's findings indicate that corrosion of reinforcing bars precedes carbonation reaching the reinforced concrete interface [25]. This prompts the hypothesis that trace amounts of CO_2 may permeate finer pores, altering the interfacial environment and leading to reinforcement corrosion. While previous studies have assessed carbonation depth or products using phenolphthalein reagent, XRD, infrared spectroscopy, and SEM, the intricate details of CO_2 transport and reaction processes remain elusive [17,19,26]. Understanding CO_2 transport in concrete pore solution is pivotal in comprehending carbonation. Thus, there is an urgent need for methodologies to visualize CO_2 transport within the intricate pores of concrete.

In recent years, the advancement of molecular dynamics (MD) simulation technology has provided effective solutions to the challenges. Hou utilized molecular dynamics to simulate water and ion adsorption and transport within CSH gel pores [15]. Subsequently, the transport processes of water and NaCl within CSH nanocone tubes were investigated [27]. Wang delved into sulfate transport within CSH pore solutions and the detachment of Ca ions at varying temperatures [28]. Tu illustrates the coupling of ions (SO_4^{2-}, Cl-, NO_2) within the concrete, revealing that sulfate ions diminish the rust-prevention effect of nitrite ions [29]. While numerous studies have employed MD methods to simulate ion and water transport in concrete pore solutions, there exists a notable gap in research concerning carbon dioxide—a major contributor to the durability deterioration of reinforced concrete systems. Given that carbon dioxide concurrently transports and reacts within concrete pores, this study employs the reactive force field (ReaxFF) molecular simulation method to explore carbon dioxide and water transport and reaction processes within concrete pores featuring diverse calcium-to-silicon ratios. As a comparative measure, the transport process of pure water within these pores is also scrutinized. The findings indicate that CO_2 and its hydration products impede solution transport within the pores. Moreover, an elevated Ca/Si ratio renders the matrix more prone to reacting with CO_2 and its hydration products, consequently expediting solution transport within the pores.

2. Modeling and Simulation Details

2.1. Model Construction

The models employed in this study were constructed based on the 11 Å tobermorite model and the realistic cement hydration model proposed by Pellenq et al. [30]. Given that carbonation significantly impacts CSH with different Ca/Si ratios—a crucial factor in carbonation studies—three CSH models with distinct Ca/Si ratios were initially constructed to analyze this effect: Ca/Si = 1, Ca/Si = 1.5, and Ca/Si = 1.5 (no bridging silica–oxygen). Among these, the Ca/Si = 1.5 (no bridging silica–oxygen) model, obtained by eliminating all bridging-site silica–oxygen from the 11 Å tobermorite model, was deemed ideal. This choice aimed to investigate whether variations in Ca-Si ratios influenced the extent of surface bridging-site silica–oxygen deletion, thereby affecting the degree of carbonation. Subsequently, the CSH with the three different Ca-Si ratios was scaled up by a factor of two along the *z*-axis direction of the unit cell. The middle part of the substrate was then removed from the (001) plane, leaving a thickness of 22.77 Å for both upper and lower substrates. The substrates were enlarged by a factor of three along the *x*-axis and a factor of 10 along the *y*-axis, resulting in parallel CSH substrates measuring 33.48 Å × 73.9 Å × 22.77 Å. Additionally, a packmol procedure introduced 6000 water molecules and 400 carbon dioxide molecules into a space of approximately 85 Å on the left side of the y-direction substrate [31]. To prevent water and ions from escaping, a vacuum layer of approximately 60 Å was retained on the right side of the y-direction substrate and on the left side of the solution. The finalized model, depicted in Figure 1, featured dimensional parameters of a = 33.48 Å, b = 218.9 Å, c = 68.31 Å, $\alpha = 90^\circ$, $\beta = 90^\circ$, and $\gamma = 90^\circ$.

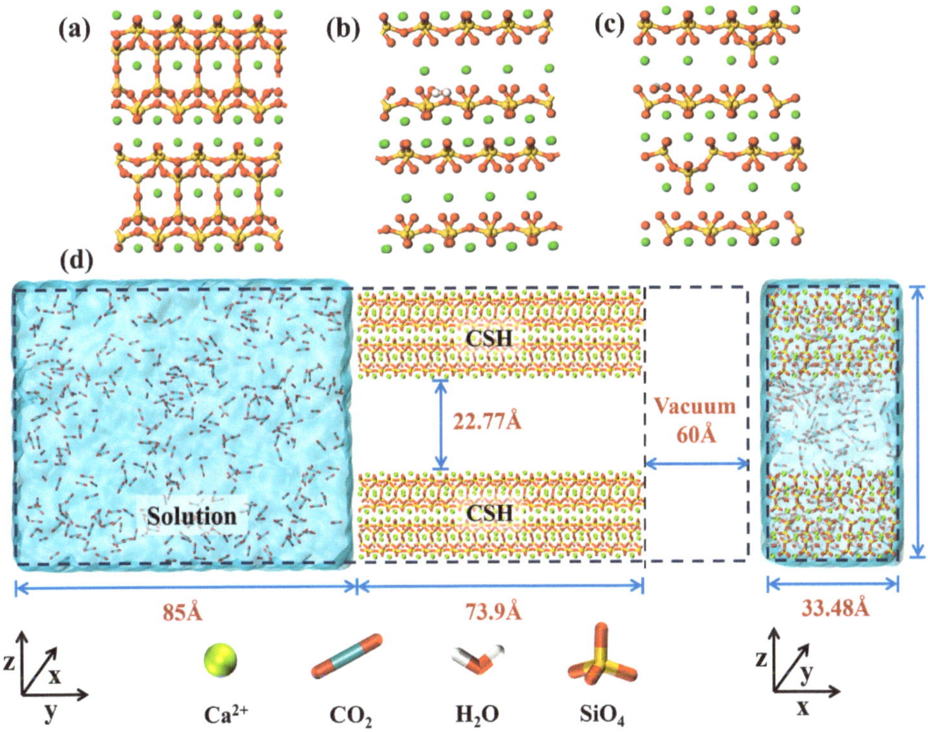

Figure 1. (**a**–**c**) show the initial unit cells for models with C/S = 1, C/S = 1.5 with non-bridging silicate–oxygen, and C/S = 1.5, respectively. After undergoing Monte Carlo simulations for water absorption to saturation, the cells were expanded, and selected atoms were removed to form pore channels. A water box containing CO_2 was then placed on the left side of the pore channel to create the C-S-H pore transport model, illustrated in (**d**).

2.2. *Force Field and Molecular Dynamics Programs*

The reactive force field (ReaxFF) is a sophisticated molecular simulation technique designed to capture a wide range of chemical reactions and interactions at the atomic level. It is particularly well suited for studying systems where bond formation and breaking occur, such as in the reaction of CO_2 with the CSH matrix in our study. In this study, we employed the reactive force field (ReaxFF) to capture atom interactions throughout the simulation. ReaxFF uses a bond-order potential function to represent the interaction between atoms. This function adapts dynamically to changes in the bonding environment, allowing for the accurate simulation of bond formation, breaking, and changes in bond order during chemical reactions. The parameters for ReaxFF are derived from empirical data and quantum mechanical calculations. Force field parameters were adopted from the works of Nabankur Dasgupta et al., Fedkin et al. [32], and Subbaraman et al. [33]. These parameters define the interactions between the atoms in the system and are essential for capturing the chemical behavior of the materials under study. The entire molecular dynamics (MD) simulation was conducted using the LAMMPS Stable Release version under the NVT ensemble. Initially, an invisible wall was placed at the entrance of the nanopore channel to prevent water molecules and ions from entering the pore during the simulation's initial phase. Subsequently, a Nosé–Hoover thermostat was employed to stabilize the system's temperature at 298 K [34]. The system was then allowed to equilibrate under the NVT ensemble for 0.5 ns. Finally, the wall at the pore entrance was removed to enable the free transport of water and ions into the pore. The simulation spanned 2 ns with a time step of 0.25 fs, and atomic trajectory frames were outputted every 1000 steps to ensure data stability for subsequent analysis. Visualization of simulation results and trajectory sections was conducted using the VMD 1.9.3 software post-simulation [35]. The reactive force field (ReaxFF) is used to model and observe the chemical reactions between CO_2 and the CSH matrix, including bond formation, reaction products, and changes in material structure.

3. Results and Discussion

3.1. Solution and Ion Transport Processes

Figure 2 illustrates the migration of an aqueous solution containing CO_2 from the solution system into CSH pores with Ca/Si = 1, Ca/Si = 1.5 (no bridging silica–oxygen), and Ca/Si = 1.5 configurations over a duration of 1500 ps. Notably, the leading interface of the solution inside the CSH nanopores adopts a distinct concave shape, and the contact angle of the liquid–solid interface measures less than 90°, indicative of the remarkable hydrophilicity of CSH. Surprisingly, Ca/Si = 1.5 (Figure 2b) exhibits the fastest transport, followed by Ca/Si = 1 (Figure 2a), while Ca/Si = 1.5 (no bridging silica–oxygen) (Figure 2c) displays the slowest transport; moreover, it was noted that at the entrance of the CSH pores, the matrix protrudes slightly, accompanied by the diffusion of silica–oxygen tetrahedra into the channels. This suggests that the transport rate of the solution within CSH pores is not solely influenced by surface structure variations due to differences in the Ca/Si ratio and the absence of bridging silicon–oxygen sites, which contributes to the instability of the CSH matrix.

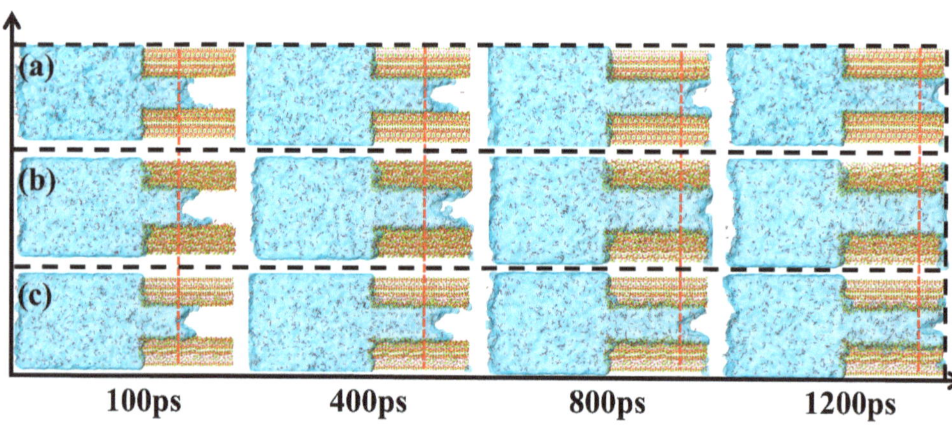

Figure 2. Snapshots of the transport of H_2O and CO_2 within the capillary channels of CSH gels in Ca/Si = 1 (**a**), Ca/Si = 1.5 (**b**), and Ca/Si = 1.5 (no bridging silica–oxygen) (**c**), the red line represents a vertical extension from the lowest point of the liquid level across the three models, allowing us to observe the differences in transmission depth.

To quantitatively assess the impact of varying calcium–silica ratios and solutions, we tracked the passage of water and ions through the channels over simulation time and determined the transport depth by measuring the distance of the intermediate level of the solution inside the pore. As depicted in Figure 3, the system with CO_2 initially enters the pore channel slightly faster than the pure water solution. However, in the latter half of the transport, the depth of transport is not as fast as that of the pure water system. This discrepancy may stem from CO_2 gas presence, which initially accelerates solution diffusion but subsequently leads to CO_2, CO_3^{2-}, or HCO_3^- occupying adsorption sites or reacting with the substrate, thereby narrowing the pore channel and slowing solution transport [27].

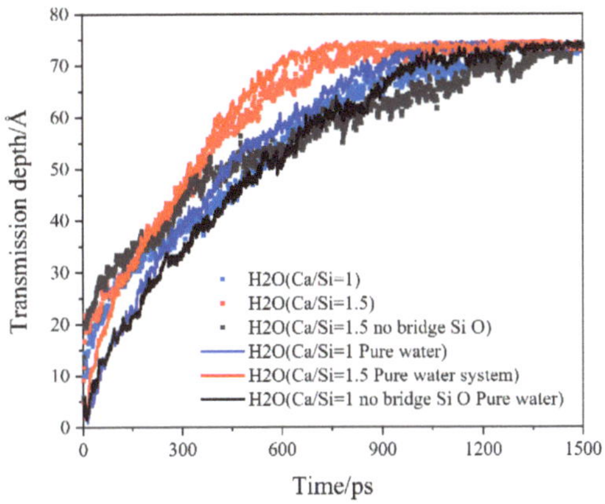

Figure 3. The depth of transport for H_2O and CO_2 within CSH nanopores under three Ca-Si ratios, with the presence of a CO_2 system.

Furthermore, we observed that the transport of CO_2, CO_3^{2-}, and HCO_3^- is contingent on the aqueous solution, with water serving as their transport medium. Figure 4 presents a comparison of different Ca/Si ratios, revealing that solution transport is fastest in the Ca/Si = 1.5 system (Figure 4b), followed by Ca/Si = 1 (Figure 4a), and Ca/Si = 1.5 (no bridging silica–oxygen) system (Figure 4b) exhibiting the slowest transport. This prompts the following question: what is the underlying mechanism driving this phenomenon? Let us delve into the specific mechanism.

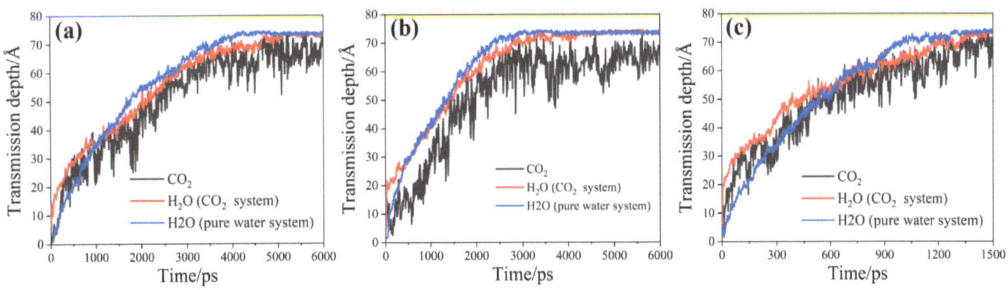

Figure 4. Depth of transport within the CSH gel pores for H_2O and CO_2 in the CO_2 system and H_2O in the pure water system, under different Ca/Si ratios: (**a**) Ca/Si = 1, (**b**) Ca/Si = 1.5, and (**c**) Ca/Si = 1.5 (no bridging silica–oxygen).

We first analyze the effect of water adsorption. The density distribution of the calculated order parameter (S_m) of water is shown in Figure 5. The order parameter is a relevant metric for assessing molecular arrangement and is calculated as follows [36,37]:

$$Sm = \frac{1}{2}(3cos^2 2\beta - 1) \quad (1)$$

where β represents the angle between the two directions of the water dipole. A S_m value close to 1 indicates high water ordering, while a value close to -0.5 suggests low ordering and a disordered state. As depicted in Figure 5, the system with a Ca/Si ratio of 1 exhibits the highest degree of water ordering, implying greater water involvement in its adsorption followed by the system with a Ca/Si ratio of 1.5. Conversely, the system with a Ca/Si ratio of 1.5 (no bridging silica–oxygen) displays the lowest degree of ordering. This discrepancy arises from the tendency of water to adsorb onto the oxygen of the silicon chain on the CSH surface, forming stronger hydrogen bonding interactions compared to Ca ions. Furthermore, oxygen at the bridging position is closer to the surface and facilitates water adsorption more readily. Consequently, differences in the bridging position of silicon–oxygen among the three systems result in varying degrees of water ordering.

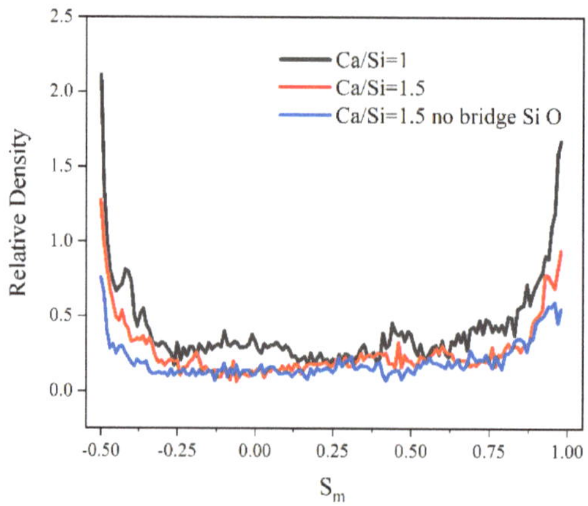

Figure 5. Order parameter of water molecules for three calcium–silica ratios (with CO_2 system).

Additionally, as depicted in Figure 6, the dipole profiles of water molecules in the three systems illustrate the distribution of water in each direction. Remarkably, the addition of CO_2 has minimal effect on water distribution. Given the hydrophilicity of CSH, water dipoles are predominantly distributed at angles of 180° and 300° to 360°.

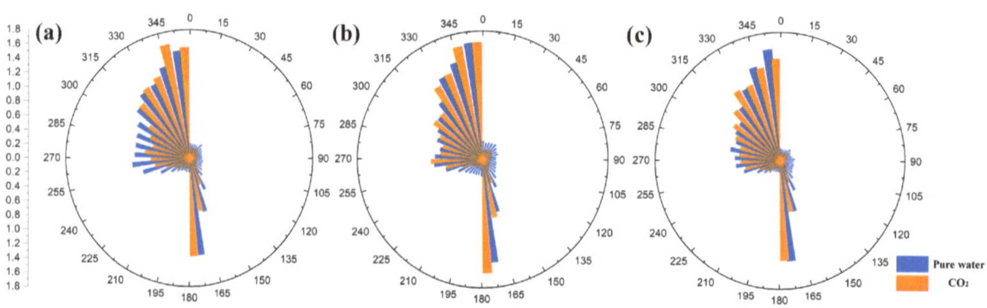

Figure 6. Dipole moment profiles of water molecules for three different systems: (**a**) Ca/Si = 1, (**b**) Ca/Si = 1.5, and (**c**) Ca/Si = 1.5 with non-bridging silica–oxygen, both in the presence of CO_2 and with pure water. These profiles illustrate the orientation and behavior of water molecules within the CSH matrix under varying Ca/Si ratios, highlighting the effect of CO_2 on the water dipole moments and structural changes in the C-S-H phases.

3.2. Distribution and Transmission Rate of H_2O and CO_2

The transport and adsorption of water and carbon dioxide can be intuitively visualized through their density distributions. Figure 7 presents the one-dimensional density distribution of water and carbon dioxide, along with silicon atoms, while Figure 8 illustrates the two-dimensional density distribution of carbon dioxide and silicon atoms.

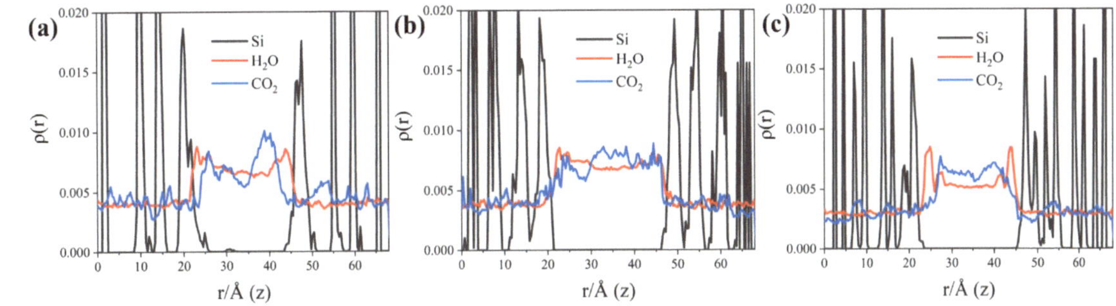

Figure 7. One-dimensional density distributions of Si atoms, H_2O, and CO_2 in the CO_2 system for different Ca/Si ratios are presented as follows: (**a**) Ca/Si = 1, (**b**) Ca/Si = 1.5, and (**c**) Ca/Si = 2.0 (excluding bridging silica–oxygen). The density distributions were extracted from the final 500 frames of the simulation, corresponding to the time range from 1.5 ns to 2.0 ns. In these distributions, the x-axis represents the z-coordinate direction, and the y-axis shows the density within the model.

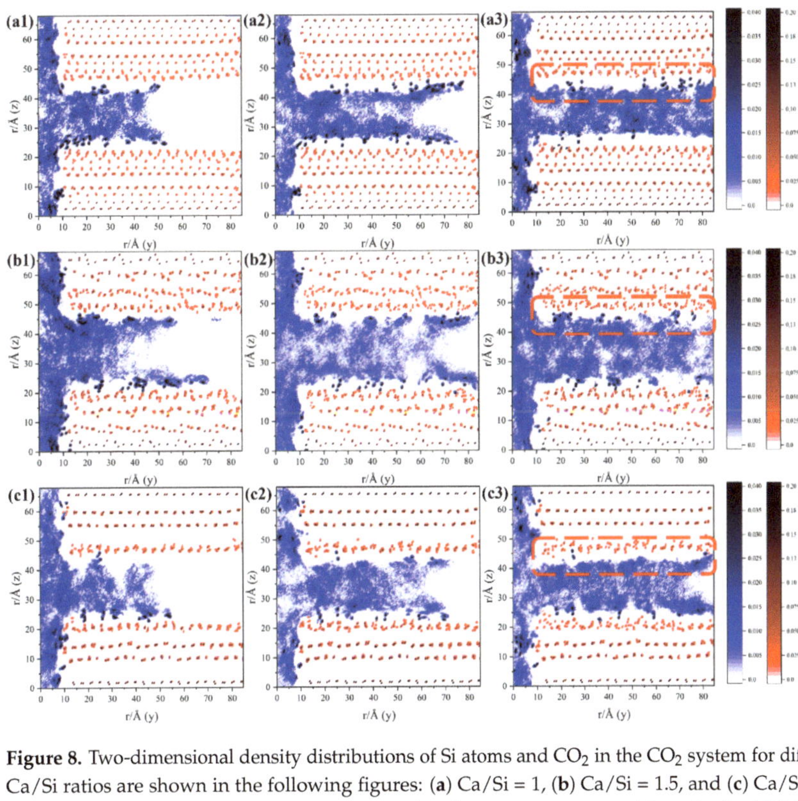

Figure 8. Two-dimensional density distributions of Si atoms and CO_2 in the CO_2 system for different Ca/Si ratios are shown in the following figures: (**a**) Ca/Si = 1, (**b**) Ca/Si = 1.5, and (**c**) Ca/Si = 2.0 (excluding bridging silica–oxygen). The density distributions were extracted from the final 500 frames of the simulation, spanning the time range from 1.5 ns to 2.0 ns. In these distributions, the x-axis represents the z-coordinate direction, and the y-axis indicates the density within the model, the red box highlights the comparison of CO_2 adsorption in CSH.

In the Ca/Si = 1 system (Figure 7a), two distinct water peaks near the surface of CSH are evident, reaffirming the highly ordered nature of water adsorption. Both Figures 7a and 8a3 indicate a noticeable gap between carbon dioxide and the CSH surface, showcasing the well-ordered distribution of Si atoms.

The distribution of Ca/Si = 1.5 (no bridging silica–oxygen) and carbon dioxide is depicted in Figure 8, illustrating the density distribution of carbon dioxide. In Ca/Si = 1.5 (no bridging silica–oxygen) (Figures 7c and 8c), the absence of bridging SiO_2 leads to water adsorption closer to the CSH surface, with slight intrusion into the matrix. Although carbon dioxide adsorption on the CSH surface still exhibits a gap, it displays a stronger peak due to the increased pore structure and adsorption sites provided by CSH without bridging SiO_2. Additionally, the matrix is more ordered in Ca/Si = 1.5 (no bridging silica–oxygen).

Contrastingly, in Ca/Si = 1.5 (Figures 7b and 8b), the matrix appears more disordered, with carbon dioxide directly adsorbed onto the CSH matrix rather than the water layer. Both water and carbon dioxide intrude on the matrix surface, resulting in an expanded transport area. This phenomenon may contribute to the faster transport rate of Ca/Si = 1.5. However, the intrusion of water and carbon dioxide inevitably leads to significant matrix damage, the effects of which are analyzed by mean square displacement (MSD).

The effect of the adsorption and stripping phenomena of CSH by fluids and CO_2 on kinetic properties can be analyzed through mean square displacement (MSD), as illustrated in Figure 9. Mean square displacement (MSD) represents the average of statistical particle displacements over time and is utilized to assess the kinematic properties of molecules and ions. The MSD is calculated as follows:

$$MSD(t) = \sum_{i=1}^{n} \frac{<|r_i(t) - r_i(0)|^2>}{n} \tag{2}$$

where *ri(t)* denotes the position of atom *i* at moment t, *ri(0)* denotes the initial position of atom *i*, and n denotes the n atoms involved in the calculation.

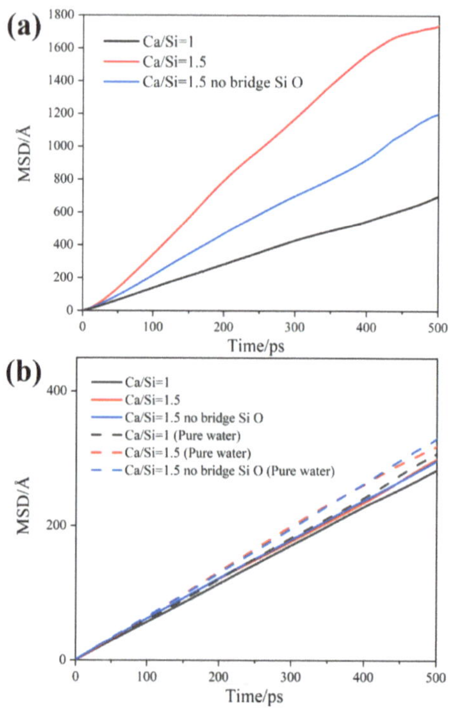

Figure 9. Mean square displacement (MSD) curves of (**a**) CO_2 and (**b**) H_2O for three different calcium-to-silicon ratios: Ca/Si = 1, Ca/Si = 1.5, and Ca/Si = 1.5 with non-bridging silica–oxygen. These curves demonstrate the mobility of CO_2 and H_2O molecules within the C-S-H pore structures, showing how varying Ca/Si ratios influence the diffusion behavior of these molecules in the CO_2 system.

In Figure 9a, the carbon dioxide transmission rates in the three models are depicted. It is notable that CO_2 in Ca/Si=1.5 exhibits the fastest transmission rate. Traditionally, in previous pore transmission studies, wider pores correlate with faster transmission rates. Therefore, as observed in Figures 7b and 8b3, CO_2 in Ca/Si = 1.5 is directly adsorbed onto the surface of the CSH, enlarging the transmission area of CO_2. This phenomenon may explain the rapid transmission rate of CO_2 in this model.

Furthermore, Ca/Si = 1.5 (no bridging silica–oxygen) demonstrates a faster transmission rate. The absence of bridging silica–oxygen widens the transmission pore of CO_2, consequently accelerating its transmission rate compared to Ca/Si = 1.

In Figure 9b, the transport velocities of water molecules in the six systems are presented. It is observed that there is not a substantial difference among all the systems, possibly due to similar water transport spaces across the three models. Additionally, it is noteworthy that water transport speeds within the same model system containing CO_2 are slower than those in pure water, implying that the presence of carbon dioxide inhibits water transport within the pores. To comprehend the cause of these phenomena, further analysis of intermolecular interactions is required.

3.3. Reasons for Transmission Differences

The main interactions between atoms are analyzed through radial distribution functions (RDF) [37–40], commonly used to represent structural properties and reflect interatomic interactions in molecular dynamics simulations. Subsequently, the stability of bonding between RDF pairs is qualitatively assessed using the time correlation function (TCF) [41–44], calculated as follows:

$$C(t) = \frac{< \delta b(t)\delta b(0) >}{< \delta b(0)\delta b(0) >} \quad (3)$$

Indeed, $\delta b(t)$ represents a binary operator. When a non-covalent bond is formed or not formed, $\delta b(t)$ equals 1 or 0, respectively. As time progresses, the interaction deteriorates, leading to a transition in TCF value from 1 to 0. Consequently, the average lifetime of various bonds can be assessed by calculating the TCF.

Figures 10 and 11 illustrate the radial distribution functions (RDF) and time correlation function (TCF) plots of calcium (Ca) on the CSH matrix with oxygen on the matrix Si (O_{Si}) and oxygen in carbon dioxide (O_C), respectively. Table 1 presents their coordination numbers.

Table 1. Coordination number of oxygen (O) and surface calcium ions (Ca) in (a) silica–oxygen tetrahedra and (b) in CO_2 and its hydration products for three calcium–silica ratios in CO_2-containing systems.

Coordination Number	O_{si}-Ca	O_C-Ca
Ca/Si = 1	0.110	0.019
Ca/Si = 1.5	0.303	0.024
Ca/Si = 1.5 (no bridging)	0.118	0.013

In Figure 10a, the RDF peaks of Ca/Si = 1 and Ca/Si = 1.5 (no bridging silica–oxygen) with O_{Si} are nearly identical, with peaks around 2.7 Å. However, the peak of Ca/Si = 1.5 (unbridged silica–oxygen) is slightly higher, and its peak position is farther from the peaks of the other two systems, at approximately 3.7 Å. Additionally, its peak is wider, indicating a higher coordination number. In Figure 10b, the RDFs of Ca and O_C for the three systems exhibit peaks at similar positions, with Ca/Si = 1 displaying the highest peak, followed by Ca/Si = 1 and Ca/Si = 1.5 (no bridging silica–oxygen) displaying the smallest peak. Analysis of the RDF and coordination number suggests that in the Ca/Si = 1.5 system, Ca distribution on the surface is wider, and more Ca is adsorbed by CO_2. This phenomenon is attributed to the intrusion of water and CO_2 on the surface of the Ca/Si = 1.5 system, weakening its Ca adsorption on the surface. Meanwhile, constant adsorption of Ca by CO_2, transported along with it, results in loosely distributed Ca on the CSH surface. Conversely, in the system without bridging silica–oxygen, stronger Ca adsorption on the surface occurs due to more available adsorption sites, reducing CO_2 competition for Ca.

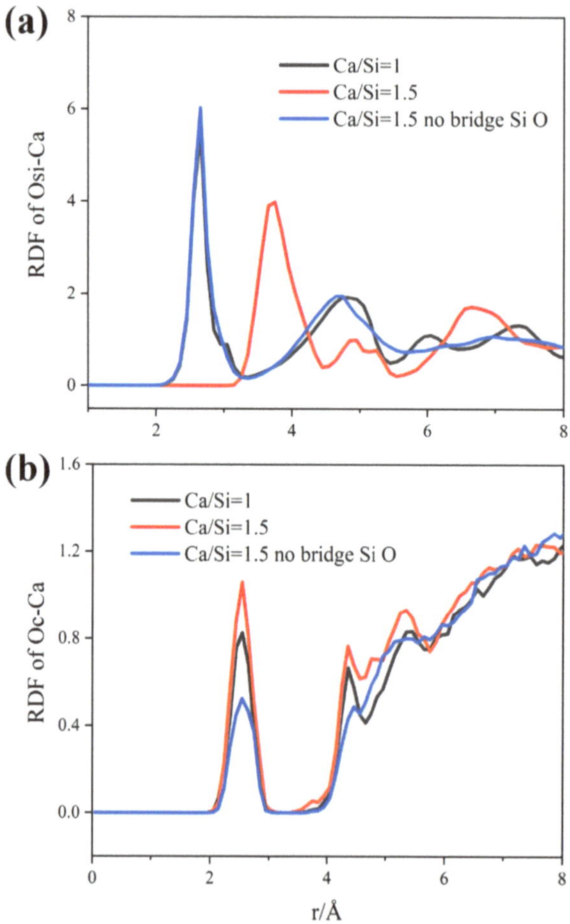

Figure 10. Radial distribution functions (RDF) of oxygen (O) and surface calcium ions (Ca) in (**a**) silica–oxygen tetrahedra and (**b**) in CO_2 and its hydration products for three calcium–silica ratios in CO_2-containing systems.

The TCF plots in Figure 11a indicate that the difference in adsorption stability of the three systems for Ca is not substantial. However, the stability of CO_2 adsorption for Ca (Figure 11b) shows significant variation, with Ca/Si = 1.5 exhibiting the worst stability. This instability contributes to the fast transport of CO_2 in the Ca/Si = 1.5 system. Further analysis of the remaining two systems will be conducted in conjunction with the BTE.

The time evolution of (non)covalent bonds (BTE) is used to represent the change in bond number with time [37,45–47]. As shown in Figure 12a, the O_{si}-Ca bond number fluctuates greatly with time in all three systems, with a tendency to oscillate periodically.

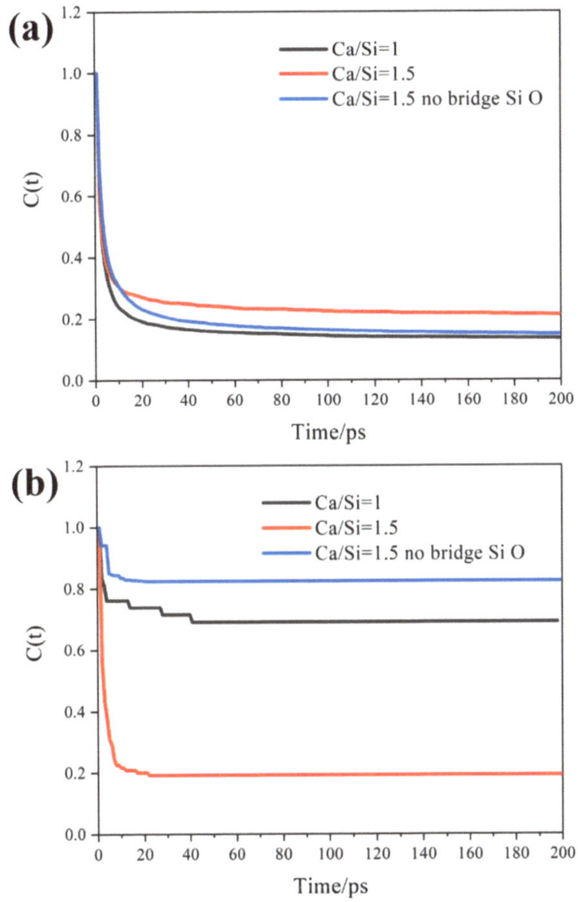

Figure 11. Time correlation function (TCF) of oxygen (O) and surface calcium ions (Ca) in (**a**) silica–oxygen tetrahedra and (**b**) CO_2 and its hydration products for systems with three different calcium-to-silica ratios: Ca/Si = 1, Ca/Si = 1.5, and Ca/Si = 1.5 with non-bridging silica–oxygen. These plots illustrate the interaction dynamics and stability of surface calcium ions with oxygen in the C-S-H matrix under CO_2 exposure.

In Figure 12b, it is evident that the O_c-Ca bond number fluctuates only in the system with Ca/Si = 1.5. This fluctuation is attributed to the constant breaking of the O_c-Ca bond, with the generation of surface Ca and the diffusion of CO_2 continuously pulling away from the surface of the CSH. This phenomenon contributes to the observed differences in the position of the RDF peaks in Figure 12a. Additionally, the Ca/Si = 1.5 (no bridging silica–oxygen) bond number shows no fluctuation at all, while Ca/Si = 1 exhibits slight fluctuation. Analysis of Table 2 reveals that the initial number of O_c-Ca bonds in Ca/Si = 1.5 (no bridging silica–oxygen) is only 4, significantly less than the 46 for Ca/Si = 1.5 and 20 for Ca/Si = 1. This finding further confirms that the absence of bridging silica–oxygen increases the adsorption sites of Ca in this system, leading to strong adsorption on the surface of CSH. However, only a small amount of Ca diffuses into the liquid to form strong adsorption with CO_2 (Figure 13c). This observation also explains why the O_c-Ca bond is most stable in Figure 12b with Ca/Si = 1.5.

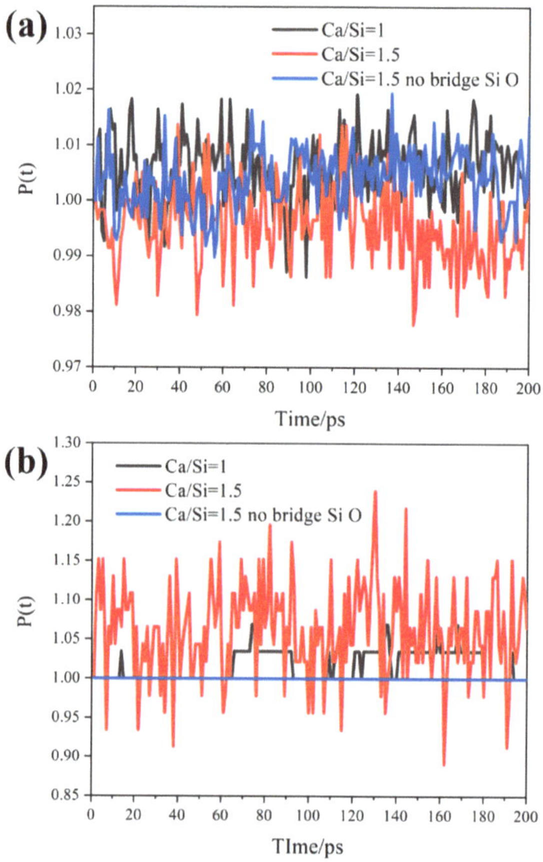

Figure 12. Binding energy (BTE) profiles of oxygen (O) and surface calcium ions (Ca) in (**a**) silica–oxygen tetrahedra and (**b**) CO_2 and its hydration products for systems with three different calcium-to-silica ratios: Ca/Si = 1, Ca/Si = 1.5, and Ca/Si = 1.5 with non-bridging silica–oxygen. These figures demonstrate the binding interactions and stability of surface calcium ions with oxygen in the C-S-H matrix when exposed to CO_2.

Table 2. Number of bonds of oxygen (O) and surface calcium ions (Ca) in (a) silica–oxygen tetrahedra and (b) in CO_2 and its hydration products for three calcium–silica ratios in CO_2-containing systems.

Number of Bonds	O_{si}-Ca	O_C-Ca
Ca/Si = 1	1095	46
Ca/Si = 1.5	980	29
Ca/Si = 1.5 (no bridging)	586	5

Through RDF analysis, it is evident that while the Ca/Si = 1 system also exhibits a stronger adsorption capacity for Ca, the amount of Ca diffused into the liquid is minimal. This Ca diffused into the solution forms a more stable bond with CO_2. Conversely, in both Ca/Si = 1.5 and Ca/Si = 1.5 (no bridging silica–oxygen), CO_2 reacts with the silica–oxygen tetrahedra (Figure 13d).

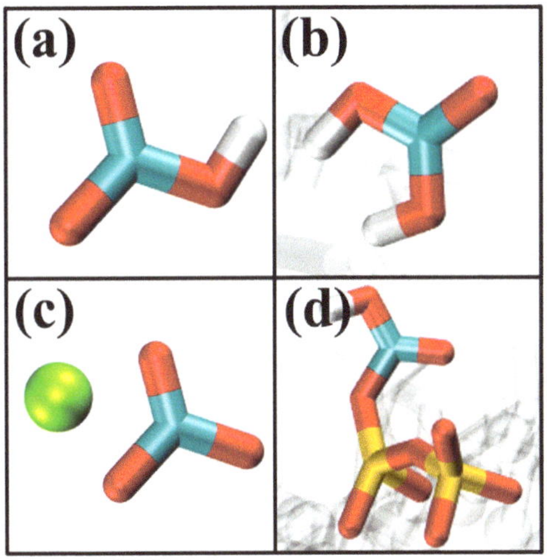

Figure 13. Reaction of bicarbonate ions (HCO_3^-) (**a**), carbonic acid (H_2CO_3) (**b**), calcium carbonate ($CaCO_3$) (**c**), and carbonic acid with CSH substrate (**d**) in three calcium-to-silica ratios within the CO_2 system, as well as Ca/Si = 1.5 and Ca/Si = 1.5 (no bridging silica–oxygen).

4. Conclusions

This study investigates the interaction of CO_2 and its hydration products with the CSH matrix, focusing on their effects on adsorption sites and transport pores. It was found that CO_2, along with its hydration products, reacts with the CSH matrix, occupying adsorption sites and narrowing transport pores, which subsequently inhibits solution transport within CSH pores.

As the Ca/Si ratio increases, CO_2 directly adsorbs onto the CSH matrix, thereby expanding the transport space and accelerating the transport rate. The adsorption of calcium ions (Ca) on the CSH surface decreases, which facilitates an increased carbonation rate due to the higher reaction of CO_2 with Ca. The continuous bonding and breaking interactions between CO_2 and its hydration products with surface Ca ions result in a broader distribution of Ca ions on the CSH surface, further enhancing the carbonation rate.

Additionally, the observed phenomena are not solely attributed to the absence of surface-bridging silica–oxygen linked to different Ca/Si ratios. Instead, it is the lack of bridging silica–oxygen that provides more adsorption sites and space for Ca ions, influencing the stability and adsorption strength of Ca. While a higher Ca/Si ratio typically leads to stronger Ca adsorption, the absence of bridging silica–oxygen adversely affects the overall structural stability.

Author Contributions: Conceptualization, P.W., Y.Z., X.W., D.H. and M.W.; methodology, P.W., Y.Z. and X.W.; data curation, W.S., C.L. and F.H.; formal analysis, C.L. and F.H.; writing—original draft preparation, W.S.; writing—review and editing, D.H. and M.W.; supervision, M.W. All authors have read and agreed to the published version of the manuscript.

Funding: M.W. and D.H. received financial support from the National Key Research and Development project 2022YFE0133800, the National Natural Science Foundation of China under grants U2006224, 52308263, 51978352, and 52178221, Shandong Province Natural Science Foundation under grant ZR2022YQ55, 2019KJG010, Shandong Provincial Education Department under grant 2019KJG010, and Qingdao Research Program 16-5-1-96-jch, 111 Project TMduracon20220022.

Data Availability Statement: The data is available by requesting from the corresponding author.

Conflicts of Interest: The authors declare no conflict of interest.

References

1. Chang, C.-F.; Chen, J.-W. The Experimental Investigation of Concrete Carbonation Depth. *Cem. Concr. Res.* **2006**, *36*, 1760–1767 [CrossRef]
2. Tang, B.; Fan, M.; Yang, Z.; Sun, Y.; Yuan, L. A Comparison Study of Aggregate Carbonation and Concrete Carbonation for the Enhancement of Recycled Aggregate Pervious Concrete. *Constr. Build. Mater.* **2023**, *371*, 130797. [CrossRef]
3. Qian, R.; Li, Q.; Fu, C.; Zhang, Y.; Wang, Y.; Jin, N.; Jin, X. Investigations on Atmospheric Carbonation Corrosion of Concrete Structure Beam Exposed to Real Marine-Environment for 7 Years. *J. Build. Eng.* **2023**, *71*, 106517. [CrossRef]
4. Kahyarian, A.; Brown, B.; Nesic, S. Electrochemistry of CO_2 Corrosion of Mild Steel: Effect of CO_2 on Iron Dissolution Reaction *Corros. Sci.* **2017**, *129*, 146–151. [CrossRef]
5. Ai, Z.; Jiang, J.; Sun, W.; Jiang, X.; Yu, B.; Wang, K.; Zhang, Z.; Song, D.; Ma, H.; Zhang, J. Enhanced Passivation of Alloy Corrosion-Resistant Steel Cr10Mo1 under Carbonation—Passive Film Formation, the Kinetics and Mechanism Analysis. *Cem. Concr. Compos.* **2018**, *92*, 178–187. [CrossRef]
6. Ming, J.; Wu, M.; Shi, J. Passive Film Modification by Concrete Carbonation: Re-Visiting a Corrosion-Resistant Steel with Cr and Mo. *Cem. Concr. Compos.* **2021**, *123*, 104178. [CrossRef]
7. Roventi, G.; Bellezze, T.; Giuliani, G.; Conti, C. Corrosion Resistance of Galvanized Steel Reinforcements in Carbonated Concrete: Effect of Wet—Dry Cycles in Tap Water and in Chloride Solution on the Passivating Layer. *Cem. Concr. Res.* **2014**, *65*, 76–84. [CrossRef]
8. Shi, J.; Geng, Z.; Zhou, X. Reinforcing Steels in Low-Carbon Mortars Subjected to Chloride Attack and Natural Carbonation: Contradictory Trends in Passivation Ability and Corrosion Resistance. *Cem. Concr. Compos.* **2024**, *152*, 105666. [CrossRef]
9. Mei, K.; He, Z.; Yi, B.; Lin, X.; Wang, J.; Wang, H.; Liu, J. Study on Electrochemical Characteristics of Reinforced Concrete Corrosion under the Action of Carbonation and Chloride. *Case Stud. Constr. Mater.* **2022**, *17*, e01351. [CrossRef]
10. Al Fuhaid, A.F.; Niaz, A. Carbonation and Corrosion Problems in Reinforced Concrete Structures. *Buildings* **2022**, *12*, 586. [CrossRef]
11. Ramirez, D.E.A.; Meira, G.R.; Quattrone, M.; John, V.M. A Review on Reinforcement Corrosion Propagation in Carbonated Concrete—Influence of Material and Environmental Characteristics. *Cem. Concr. Compos.* **2023**, *140*, 105085. [CrossRef]
12. Angst, U.; Moro, F.; Geiker, M.; Kessler, S.; Beushausen, H.; Andrade, C.; Lahdensivu, J.; Köliö, A.; Imamoto, K.; von Greve-Dierfeld, S.; et al. Corrosion of Steel in Carbonated Concrete: Mechanisms, Practical Experience, and Research Priorities—A Critical Review by RILEM TC 281-CCC. *RILEM Tech. Lett.* **2020**, *5*, 85–100. [CrossRef]
13. Rusheng, Q.; Yunsheng, Z.; Yu, Z.; Chuanqing, F.; Cheng, L.; Lin, Y.; Guojian, L. Various Gas Transport Properties in Concrete Considering Transporting Mechanisms and Testing Methods-A Review. *Constr. Build. Mater.* **2023**, *389*, 131636. [CrossRef]
14. Holland, B.; Alapati, P.; Kurtis, K.E.; Kahn, L. Effect of Different Concrete Materials on the Corrosion of the Embedded Reinforcing Steel. In *Corrosion of Steel in Concrete Structures*; Elsevier: Amsterdam, The Netherlands, 2023; pp. 199–218.
15. Hou, D.; Li, Z. Molecular Dynamics Study of Water and Ions Transport in Nano-Pore of Layered Structure: A Case Study of Tobermorite. *Microporous Mesoporous Mater.* **2014**, *195*, 9–20. [CrossRef]
16. Kaddah, F.; Ranaivomanana, H.; Amiri, O.; Rozière, E. Accelerated Carbonation of Recycled Concrete Aggregates: Investigation on the Microstructure and Transport Properties at Cement Paste and Mortar Scales. *J. CO2 Util.* **2022**, *57*, 101885. [CrossRef]
17. Auroy, M.; Poyet, S.; Le Bescop, P.; Torrenti, J.-M.; Charpentier, T.; Moskura, M.; Bourbon, X. Impact of Carbonation on Unsaturated Water Transport Properties of Cement-Based Materials. *Cem. Concr. Res.* **2015**, *74*, 44–58. [CrossRef]
18. Phung, Q.T.; Maes, N.; Jacques, D.; De Schutter, G.; Ye, G.; Perko, J. Modelling the Carbonation of Cement Pastes under a CO_2 Pressure Gradient Considering Both Diffusive and Convective Transport. *Constr. Build. Mater.* **2016**, *114*, 333–351. [CrossRef]
19. Kari, O.-P.; Puttonen, J.; Skantz, E. Reactive Transport Modelling of Long-Term Carbonation. *Cem. Concr. Compos.* **2014**, *52*, 42–53. [CrossRef]
20. Sharmilan, S.; Stang, H.; Michel, A. A Multi-Species Reactive Transport Model Based on Gas-Ion-Solid Phase Interaction for the Carbonation of Cement-Based Materials. *Cem. Concr. Res.* **2024**, *175*, 107349. [CrossRef]
21. Xie, M.; Dangla, P.; Li, K. Reactive Transport Modelling of Concurrent Chloride Ingress and Carbonation in Concrete. *Mater. Struct.* **2021**, *54*, 177. [CrossRef]
22. Wang, X.; Kong, L.; Zhao, W.; Liu, Y. Chloride Transport Resistance of Alkali-Activated Concrete Exposed to Combined Chloride, Sulfate and Carbonation Environment. *Constr. Build. Mater.* **2023**, *367*, 130353. [CrossRef]
23. Georget, F.; Soja, W.; Scrivener, K.L. Characteristic Lengths of the Carbonation Front in Naturally Carbonated Cement Pastes: Implications for Reactive Transport Models. *Cem. Concr. Res.* **2020**, *134*, 106080. [CrossRef]
24. Sevelsted, T.F.; Skibsted, J. Carbonation of C–S–H and C–A–S–H Samples Studied by 13 C, 27 Al and 29 Si MAS NMR Spectroscopy. *Cem. Concr. Res.* **2015**, *71*, 56–65. [CrossRef]
25. Mi, T.; Li, Y.; Liu, W.; Dong, Z.; Gong, Q.; Min, C.; Xing, F.; Wang, Y.; Chu, S.H. The Effect of Carbonation on Chloride Redistribution and Corrosion of Steel Reinforcement. *Constr. Build. Mater.* **2023**, *363*, 129641. [CrossRef]
26. Bui, H.; Delattre, F.; Levacher, D. Experimental Methods to Evaluate the Carbonation Degree in Concrete—State of the Art Review. *Appl. Sci.* **2023**, *13*, 2533. [CrossRef]
27. Hou, D.; Zheng, H.; Wang, P.; Wan, X.; Yin, B.; Wang, M.; Zhang, J. Molecular Dynamics Study on Sodium Chloride Solution Transport through the Calcium-Silicate-Hydrate Nanocone Channel. *Constr. Build. Mater.* **2022**, *342*, 128068. [CrossRef]

28. Wang, F.; Zhang, Y.; Jiang, J.; Yin, B.; Li, Z. Effect of Temperature on the Capillary Transport of Sodium Sulfate Solution in Calcium Silicate Hydrate Nanopore: A Molecular Dynamics Study. *Constr. Build. Mater.* **2020**, *231*, 117111. [CrossRef]
29. Tu, Y.; Cao, J.; Wen, R.; Shi, P.; Yuan, L.; Ji, Y.; Das, O.; Försth, M.; Sas, G.; Elfgren, L. Molecular Dynamics Simulation Study of the Transport of Pairwise Coupled Ions Confined in C-S-H Gel Nanopores. *Constr. Build. Mater.* **2022**, *318*, 126172. [CrossRef]
30. Pellenq, R.J.-M.; Kushima, A.; Shahsavari, R.; Van Vliet, K.J.; Buehler, M.J.; Yip, S.; Ulm, F.-J. A Realistic Molecular Model of Cement Hydrates. *Proc. Natl. Acad. Sci. USA* **2009**, *106*, 16102–16107. [CrossRef]
31. Martínez, L.; Andrade, R.; Birgin, E.G.; Martínez, J.M. PACKMOL: A Package for Building Initial Configurations for Molecular Dynamics Simulations. *J. Comput. Chem.* **2009**, *30*, 2157–2164. [CrossRef]
32. Fedkin, M.V.; Shin, Y.K.; Dasgupta, N.; Yeon, J.; Zhang, W.; Van Duin, D.; Van Duin, A.C.T.; Mori, K.; Fujiwara, A.; Machida, M.; et al. Development of the ReaxFF Methodology for Electrolyte—Water Systems. *J. Phys. Chem. A* **2019**, *123*, 2125–2141. [CrossRef]
33. Subbaraman, R.; Deshmukh, S.A.; Sankaranarayanan, S.K.R.S. Atomistic Insights into Early Stage Oxidation and Nanoscale Oxide Growth on Fe (100), Fe (111) and Fe (110) Surfaces. *J. Phys. Chem. C* **2013**, *117*, 5195–5207. [CrossRef]
34. Nosé, S. A Unified Formulation of the Constant Temperature Molecular Dynamics Methods. *J. Chem. Phys.* **1984**, *81*, 511–519. [CrossRef]
35. William, H. VMD-Visual Molecular Dynamics. *J. Mol. Graph.* **1996**, *14*, 33–38.
36. Hou, D.; Zhang, K.; Hong, F.; Wu, S.; Wang, Z.; Li, M.; Wang, M. The Corrosion Deterioration of Reinforced Passivation Film: The Impact of Defects. *Appl. Surf. Sci.* **2022**, *582*, 152408. [CrossRef]
37. Wang, M.; Zhang, K.; Ji, X.; Wang, P.; Ma, H.; Zhang, J.; Hou, D. Molecular Insight into the Fluidity of Cement Pastes: Nano-Boundary Lubrication of Cementitious Materials. *Constr. Build. Mater.* **2022**, *316*, 125800. [CrossRef]
38. Kirkwood, J.G.; Boggs, E.M. The Radial Distribution Function in Liquids. *J. Chem. Phys.* **1942**, *10*, 394–402. [CrossRef]
39. Zimm, B.H. The Scattering of Light and the Radial Distribution Function of High Polymer Solutions. *J. Chem. Phys.* **1948**, *16*, 1093–1099. [CrossRef]
40. Hou, D.; Wang, M.; Sun, H.; Wang, P.; Zhang, J.; Wang, M. Structural Deterioration of Calcium Aluminosilicate Hydrate by Sulfate Attack: Atomic Processes of Al-Si Chain Breaking. *Constr. Build. Mater.* **2023**, *401*, 132867. [CrossRef]
41. Hong, F.; Yu, S.; Hou, D.; Li, Z.; Sun, H.; Wang, P.; Wang, M. Study on the Mechanical Properties, Gelling Products and Alkalization Process of Alkali-Activated Metakaolin: From Experiment to Molecular Dynamics Simulation. *J. Build. Eng.* **2023**, *79*, 107705. [CrossRef]
42. Hou, D.; Yu, J.; Liu, Q.; Dong, B.; Wang, X.; Wang, P.; Wang, M. Nanoscale Insight on the Epoxy-Cement Interface in Salt Solution: A Molecular Dynamics Study. *Appl. Surf. Sci.* **2020**, *509*, 145322. [CrossRef]
43. Wang, M.; Sun, H.; Zhou, X.; Wang, P.; Zhang, Y.; Wang, X.; Zhang, X.; Hou, D.; Wang, M. Atomistic Insights into the Deposition of Corrosion Products on the Surfaces of Steels and Passivation Films. *Langmuir* **2023**, *39*, 6812–6822. [CrossRef] [PubMed]
44. Hong, F.; Wang, M.; Dong, B.; Diao, X.; Zhang, X.; Pang, K.; Zhang, Y.; Hou, D. Molecular Insight into the Pozzolanic Reaction of Metakaolin and Calcium Hydroxide. *Langmuir* **2023**, *39*, 3601–3609. [CrossRef] [PubMed]
45. Xin, L.; Ji, X.; Yang, X.; Hong, F.; Hou, D.; Wang, M. Nanoscale Insight into Fluidity Improvement of Supplementary Cementitious Materials on Concrete: The Nano-Lubrication Process. *J. Build. Eng.* **2024**, *82*, 108193. [CrossRef]
46. Wang, M.; Ji, X.; Manzano, H.; Hou, D.; Li, Z. Unlocking Cementitious Performance: Nano-Lubrication via Polycarboxylate Superplasticizers. *J. Am. Ceram. Soc.* **2024**, *107*, 3055–3067. [CrossRef]
47. Wang, M.; Sun, H.; Zhou, X.; Wang, P.; Li, Z.; Hou, D. Surface Engineering of Migratory Corrosion Inhibitors: Controlling the Wettability of Calcium Silicate Hydrate in the Nanoscale. *Langmuir* **2023**, *39*, 17110–17121. [CrossRef]

buildings

MDPI

Article

Monitoring Early-Stage Evolution of Free Water Content in Alkali-Activated Slag Systems by Using ^{1}H Low-Field NMR

An Guo [1,2], Fanyuan Mu [1,2], Ting Zhang [3], Jiandong Wu [4], Zhenping Sun [1,2,*] and Jingbin Yang [1,2,*]

[1] Key Laboratory of Advanced Civil Engineering Materials, Ministry of Education, Tongji University, Shanghai 201804, China; 2331521@tongji.edu.cn (A.G.); 1810837@tongji.edu.cn (F.M.)
[2] School of Materials Science and Engineering, Tongji University, Shanghai 201804, China
[3] College of Civil Engineering, Zhejiang University of Technology, Hangzhou 310023, China; tzhang@zjut.edu.cn
[4] Shandong Provincial Communications Planning and Design Institute Group Co., Ltd., Jinan 250101, China; jdwu_mail@163.com
* Correspondence: szhp@tongji.edu.cn (Z.S.); jingbinyang@tongji.edu.cn (J.Y.)

Abstract: In the present study, the evolution of free water content in five different alkali-activated slag (AAS) systems was continuously monitored and compared using ^{1}H low-field NMR. The alkali activators used were waterglass solutions with three different moduli (1.2, 1.4, and 1.6), sodium hydroxide solution, and sodium sulfate solution. The findings reveal that the type of activator significantly affected the dynamic changes in the relative free water content. Notably, an increase in free water content was observed in the early stages of hydration of all AAS systems except for those activated by sodium sulfate solution. Additionally, this study investigated the relationship between changes in free water content and hydration heat in the AAS systems, dividing the initial 24 h of AAS hydration into three stages. The results demonstrate that free water can serve as an effective probe for monitoring the hydration process in fresh AAS pastes, offering valuable insights alongside traditional thermal signals.

Keywords: alkali-activated slag; free water content; hydration heat; ^{1}H low-field NMR

Citation: Guo, A.; Mu, F.; Zhang, T.; Wu, J.; Sun, Z.; Yang, J. Monitoring Early-Stage Evolution of Free Water Content in Alkali-Activated Slag Systems by Using ^{1}H Low-Field NMR. *Buildings* **2024**, *14*, 3079. https://doi.org/10.3390/buildings14103079

Academic Editor: Geo Paul

Received: 23 August 2024
Revised: 23 September 2024
Accepted: 24 September 2024
Published: 26 September 2024

1. Introduction

Alkali-activated slag (AAS) is recognized as a sustainable cementitious material and has been proposed as one of the viable substitutes for Portland cement in recent decades. Its use has the potential to significantly reduce energy consumption and carbon emissions associated with clinker production [1–4]. However, certain performance limitations of AAS, such as very short setting time [5,6] and high shrinkage [7,8], have constrained its widespread adoption. To enhance the utilization of AAS, many researchers have employed various techniques to investigate its hydration mechanisms and microstructure. These studies aim to understand the changes in macroscopic properties and identify potential modification methods [9]. However, due to the diverse sources of raw materials for AAS and variations in alkali activators and precursor properties [10,11], the hydration mechanism of AAS remains not fully understood.

In recent years, the ^{1}H low-field NMR technique has become extensively used in the research of cement-based materials. Water, a fundamental component in all cementitious systems, undergoes various transformations during hydration, shifting from free water to chemically bound water, physically absorbed water, and pore water. By correlating these transitions with hydration kinetics, the hydration process can be analyzed through the proton signals of water molecules in their different binding states [12]. Liu et al. [13] quantitatively assessed the hydration degree of cement pastes with water–cement ratios of 0.25, 0.35, and 0.45 under various curing conditions using ^{1}H low-field NMR to analyze transverse relaxation time (T_2) distributions. Yang et al. [14] employed ^{1}H low-field NMR to calculate the saturation degree and pore structure of the cement paste under different saturation methods. In our previous study, the liquid absorption-release behavior of SAPs in cement pastes and AAS pastes was compared using ^{1}H low-field NMR [15]. The influence of the pre-absorbed water amount and the water–cement ratios of paste on the water absorption-release behavior of SAPs in fresh cement paste was investigated by ^{1}H low-field NMR [16]. In addition to cement-based materials, the ^{1}H low-field NMR technique has gradually been applied to the study of alkali-activated materials (AAM) systems. Liang et al. [17] used ^{1}H low-field NMR to monitor the evolution of the state and relative contents of the water in metakaolin-based geopolymers prepared from waterglass activators of different moduli with curing time. Jiang et al. [18] used ^{1}H low-field NMR to monitor the AAS at various ages and estimated the average gel pores size by calculating the weighted average of transverse relaxation time. Zhang et al. [19] employed ^{1}H low-field NMR to study the hydration process of waterglass-activated fly ash-slag pastes, finding that the paste activated with low modulus showed a higher degree of hydration at an early age. Compared to cement-based materials, AAM systems are highly complex, with their hydration processes significantly influenced by the composition of the cementitious materials and the type of activators used [20,21]. Currently, there is a lack of study on the application of ^{1}H low-field NMR in investigating the early hydration process of AAM. Addressing this gap is essential for enhancing the understanding and comprehension of AAM hydration mechanisms.

As one of the representative systems of AAM, the early hydration process of AAS has garnered significant attention. Gebregziabiher et al. [22] used an in situ isothermal calorimetric method to characterize the early reaction kinetics and microstructure development of AAS pastes. Cao et al. [23] performed in situ monitoring of the early reaction process of AAS pastes using an ultrasonic monitoring system. In our previous study, the volume change in AAS pastes was measured by helium pycnometry, and the relationship between the volume change in AAS pastes and early hydration was analyzed in conjunction with the hydration exotherm [24]. In this study, the evolution of free water content during the early hydration process of the different AAS systems was continuously and comparatively monitored using ^{1}H low-field NMR with water as a probe. Additionally, the relationship between the changes in free water content and hydration heat in various AAS systems was explored.

2. Materials and Methods

2.1. Materials

In this study, ground granulated blast furnace slag (GGBFS) with a d_{50} of 10.9 μm was used as a precursor. Its chemical composition (determined by X-ray fluorescence) is given in Table 1. There are 5 different AAS systems, and their alkali activators are waterglass (Na_2SiO_3) solution with different moduli (SiO_2/Na_2O = 1.2, 1.4, and 1.6), sodium hydroxide (NaOH) solution, and sodium sulfate (Na_2SO_4) solution.

Table 1. Chemical composition of GGBFS (wt.%).

Oxides	CaO	SiO_2	Al_2O_3	MgO	SO_3	Na_2O	Others
Content	34.06	33.80	17.67	9.56	2.38	0.71	1.82

2.2. Mixture Proportions and Mixing Procedures

The alkali equivalent (W_{Na2O}/W_{slag}) for all AAS pastes was 5%, and the water-GGBFS ratio was 0.40. GGBFS was mixed with the prepared alkali activator solutions for 1 min. The codes WG1.2, WG1.4, WG1.6, SH, and SS indicate alkali activator was waterglass with a modulus of 1.2, 1.4, and 1.6, NaOH solution, and Na_2SO_4 solution, respectively.

2.3. H Low-Field NMR

An amount of 20 g of the fresh AAS pastes was poured into a quartz glass vial and sealed with polytetrafluoroethylene film for ^1H low-field NMR testing. The vial was placed into the sample chamber of the ^1H low-field NMR instrument with the chamber cap secured for continuous sampling, with a sampling interval of 5 min and a total duration of 24 h, by which time the free water content change in the early stage of the AAS systems had largely stabilized. Figure 1 shows the ^1H low-field NMR instrument. The magnetic field and radiofrequency (RF) coil of the 1H low-field NMR instrument (PQ001, Niumag China) were 0.5 T and 25 mm, respectively. The CPMG (Carr–Purcell–Meiboom–Gill) sequence was used to measure the transverse relaxation time (T_2) of the paste. Then, the T_2 relaxation curves were fitted to a multi-exponential curve using the inverse Laplace transform algorithm. The echo time ($\tau 1 = 0.302$ ms) and the number of scans (NS = 4) were kept constant throughout the testing.

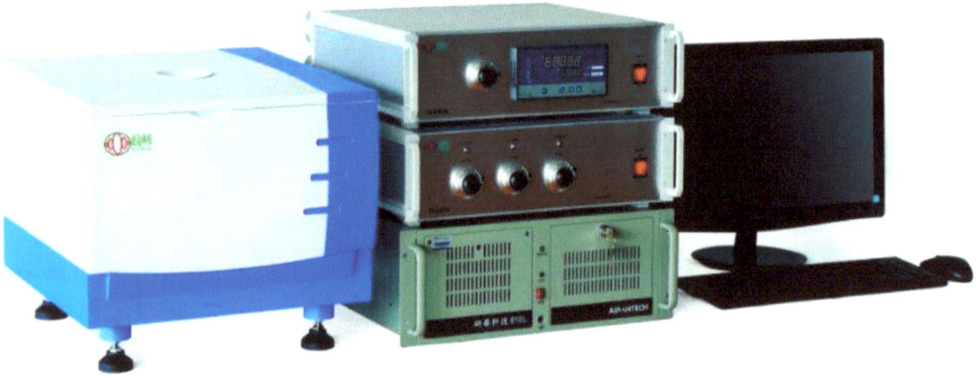

Figure 1. ^1H low-field NMR instrument.

2.4. Isothermal Calorimetry

The hydration temperature rise of AAS pastes was measured using isothermal calorimetry in accordance with GB12959-2008 [25]. The exothermic of these paste samples was continuously recorded over a 24 h hydration period.

3. Results and Discussions

3.1. Transverse Relaxation Time (T_2) Distribution of AAS Pastes

Figure 2 shows the T_2 distribution of AAS pastes at 5 min, 15 min, 50 min, and 24 h after mixing. T_2 is related to the size of the pores where free water is located; the smaller the T_2, the smaller the corresponding pore size. Therefore, it is possible to distinguish water in different pore sizes according to the T_2 distribution. The T_2 values range from short to long, corresponding to interlayer and gel water (0.01~1 ms), capillary water (1~100 ms), and surface water (100~10,000 ms). For all samples, as hydration time increases, the T_2 peak gradually shifts towards shorter relaxation times, indicating the gradual refinement of the pore space in the pastes. The T_2 peak area reflects the relative content of free water—the larger the area, the higher the free water content. Thus, changes in the free water content in AAS pastes can be analyzed based on the changes in the total T_2 peak area.

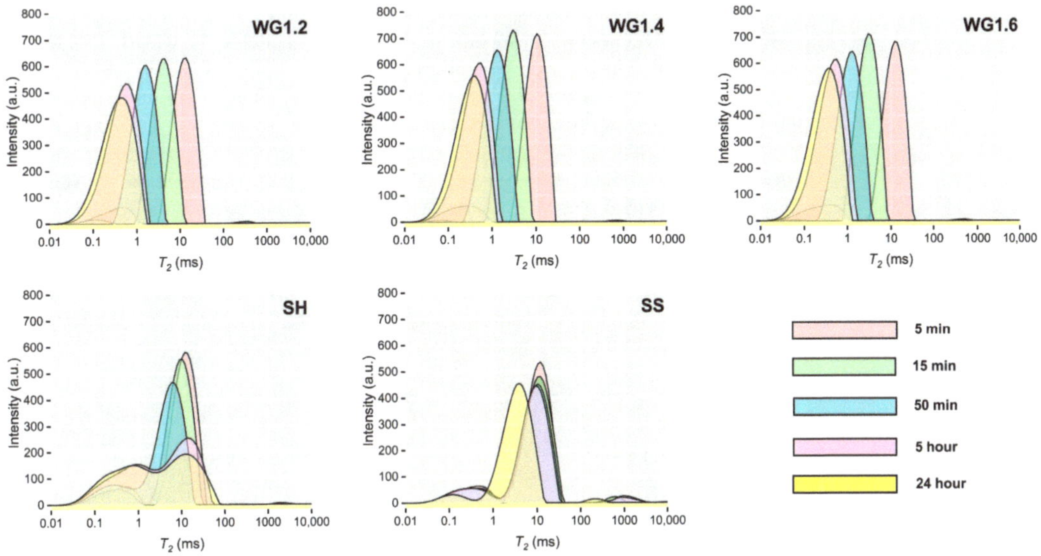

Figure 2. T_2 distribution of AAS pastes.

3.2. Total T_2 Peak Area and Relative Free Water Content

To clearly illustrate the changes in relative free water content in different AAS pastes, Figure 3 presents the total T_2 peak area per unit weight of AAS pastes over time. The free water content in all AAS pastes decreases rapidly during the first ten minutes, likely due to the wetting and dissolution of GGBFS particles and the initial formation of C-S-H [22,23,26]. In the subsequent 2~4 h, the free water content in the three WG systems begins to increase rapidly, even surpassing the initial levels. An increase in free water content was also observed in the SH system, although the increase was significantly less than that in the WG systems. However, no increase in free water content was noted in the SS system.

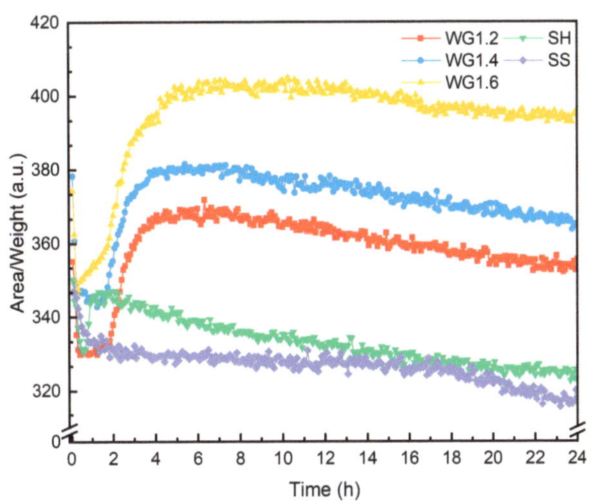

Figure 3. Total T_2 peak area per unit weight of AAS pastes.

For WG, the predominant silicate monomers in solutions are $Si(OH)_3O^-$ and $Si(OH)_2O_2^{2-}$ [27–30], and a small amount of $Si(OH)_4$ [28,29]. When the waterglass solution interacts with GGBFS particles, its high pH environment breaks the Ca-O, Mg-O, Al-O, and Si-O bonds on the surface of the GGBFS particles. This leads to the release of Ca, Mg, Al, and Si into the pore solution in the forms of Ca^{2+}, $Ca(OH)^{+}$, Mg^{2+}, $Al(OH)_4^-$, $Si(OH)_3O^-$, and $Si(OH)_2O_2^{2-}$ (Equation (1)) [27,30,31]. These introduced ions disrupt the electrostatic repulsion between the original $Si(OH)_3O^-$ and $Si(OH)_2O_2^{2-}$ in the pore solution, thereby facilitating the aggregation of the silicate structure and leading to dehydration condensation and the release of water (Equation (2)) [32]. Simultaneously, Al in the form of $Al(OH)_4^-$ formed a complex with Si simplified to silicate monomers $Si(OH)_{4-y}O_y^{y-}$ by a condensation reaction (Equation (3)) [31]. The above two processes are the primary sources of free water generation. The free water content increases with the elevation of modulus, which can be attributed to the fact that when the alkali equivalent for all AAS pastes is constant, a higher modulus introduces more $Si(OH)_3O^-$ and $Si(OH)_2O_2^{2-}$ into the system, thereby facilitating the aforementioned reactions.

$$\mathrm{GGBFS} \rightarrow \mathrm{Ca(OH)}_x^{(2-x)+} + \mathrm{Mg}^{2+} + \mathrm{Al(OH)}_4^- + \mathrm{Si(OH)}_{4-y}\mathrm{O}_y^{y-}\ \mathrm{x} = 0{,}1\ \mathrm{y} = 1{,}2 \quad (1)$$

$$2\mathrm{n} \equiv \mathrm{Si} - \mathrm{OH} \rightarrow \mathrm{n} \equiv \mathrm{Si} - \mathrm{O} - \mathrm{Si} \equiv + \mathrm{nH_2O} \quad (2)$$

$$\mathrm{Al(OH)}_4^- + \mathrm{Si(OH)}_{4-y}\mathrm{O}_y^{y-} \rightarrow \mathrm{(OH)_3Al{-}OSi(OH)}_{3-y}\mathrm{O}_y^{(y+1)-} + \mathrm{H_2O} \mathrm{y} = 1{,}2 \quad (3)$$

For SH, sodium hydroxide creates a high pH environment that breaks the Ca-O, Mg-O, Al-O, and Si-O bonds on the surface of the GGBFS particles. However, in these systems, the presence of $Al(OH)_4^-$, $Si(OH)_3O^-$, and $Si(OH)_2O_2^{2-}$ arises solely from the dissolution of GGBFS, without additional introduction of these species. Consequently, the consideration reaction is limited, producing much less free water compared to the WG system.

For SS, the activator does not contain $Si(OH)_3O^-$ and $Si(OH)_2O_2^{2-}$, and the weak alkalinity of this system cannot break the Ca-O, Mg-O, Al-O, and Si-O bonds on the surface of the GGBFS particles [27,33]. Therefore, it does not cause a significant increase in free water content.

3.3. *Evolution of Relative Free Water Content and Exothermic Hydration*

Figure 4 presents the comparative curves of free water relative content and hydration temperature rise for WG1.2, WG1.4, and WG1.6. To validate the effectiveness of the ^1H low-field NMR technique in characterizing the hydration process of AAS pastes, both the free water content change curves and hydration temperature rise curves were fitted for different stages (fitting parameters are indicated in Tables 2 and 3). In the hydration temperature rise curves, the points T_1, T_2, and T_3 represent the zero points of the first derivative, indicating the peak moments of hydration temperature rise. In the ^1H low-field NMR curves, $T_{1'}$ marks the zero point of the first derivative, corresponding to the moment of minimum free water content. The $T_{2'}$ and $T_{3'}$ are the zeros of the second derivative, signifying the turning points in the rate of change in free water content.

Table 2. Polynomial fitting parameters of free water content curves.

Line	Intercept	B1	B2	B3	B4	Adjust R-Square
WG1.2-T1′	355.85957	−92.24463	109.31328	−53.5594	9.81192	0.90438
WG1.2-T2′	660.8592	−573.7087	350.52311	−88.11199	7.95984	0.99458
WG1.2-T3′	232.71932	84.8228	−18.95517	1.72678	−0.04988	0.98493
WG1.4-T1′	357.707	−21.07717	−1.95053	10.66271	−2.36412	0.96426
WG1.4-T2′	7199.57682	−10,297.61	5766.71376	−1423.75936	130.95654	0.999157
WG1.4-T3′	234.53892	106.46055	−29.40542	3.61868	−0.16683	0.99389
WG1.6-T1′	-	-	-	-	-	-
WG1.6-T2′	340.47114	34.03166	−39.23835	20.97967	−3.29253	0.99491
WG1.6-T3′	241.07024	111.56595	−31.58026	4.18513	−0.21238	0.98795

Note: Equation: y = Intercept + B1 × x + B2 × x2 + B3 × x3 + B4 × x4.

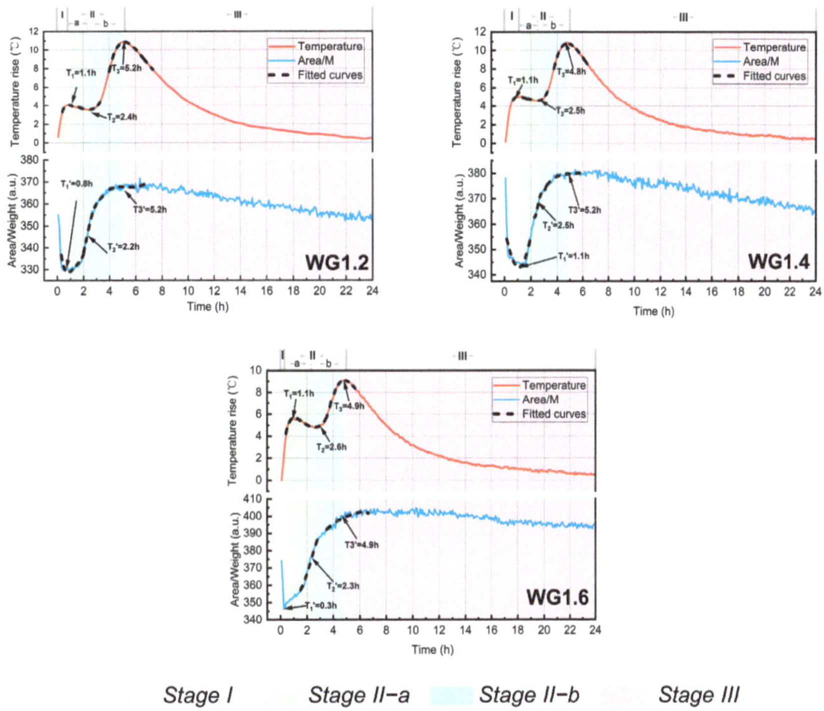

Figure 4. Free water content and hydration temperature rise of WG system paste.

Table 3. Polynomial fitting parameters of hydration temperature rise curves.

Line	Intercept	B1	B2	B3	B4	B5	Adjust R-Square
WG1.2-T1	−0.68378	16.27285	−19.78803	10.0948	−1.86767	0	0.99951
WG1.2-T2	10.77092	−11.5658	7.52827	−2.3281	0.28242	0	0.99453
WG1.2-T3	−169.2431	107.5423	−23.50679	2.23757	−0.07947	0	0.99839
WG1.4-T1	0.38426	12.72088	−12.41088	5.1256	−0.7771	0	0.91368
WG1.4-T2	−0.33207	6.90352	−2.51546	−0.0326	0.10002	0	0.97141
WG1.4-T3	−140.0046	94.63683	−21.59335	2.12849	−0.07791	0	0.99572
WG1.6-T1	0.11221	14.984	−14.73204	6.62838	−1.45539	0.12918	0.97498
WG1.6-T2	0.11221	14.984	−14.73204	6.62838	−1.45539	0.12918	0.97498
WG1.6-T3	−80.70401	46.00336	−7.13968	0.22497	0.01618	0	0.98267

Note: Equation: y = Intercept + B1 × x + B2 × x2 + B3 × x3 + B4 × x4 + B5 × x5.

Based on these two curves, the hydration process of WG is divided into three stages: the dissociation stage (Stage I), the transition stage (Stage II), and the diffusion-controlled stage (Stage III). The transition stage is further subdivided into the deceleration stage (Stage II-a) and the acceleration stage (Stage II-b).

Stage I: In the hydration temperature rise curve, an initial peak appears during this stage. In the free water content curve, a rapid decrease in free water content is observed. This decrease is attributed to the early wetting, dissolution of GGBFS particles, and the initial formation of C-S-H [22,23,26]. When GGFSS dissolves in an alkali environment, the bond energies of Si-O and Al-O are higher than those of Ca-O and Mg-O, causing Ca and Mg to dissolve first. As a result, a low Si-rich layer consisting of C-S-H, C-A-H, and C-A-S-H with a low Ca/Si ratio quickly forms on the surface of GGBFS, as shown in Figure 5a [27].

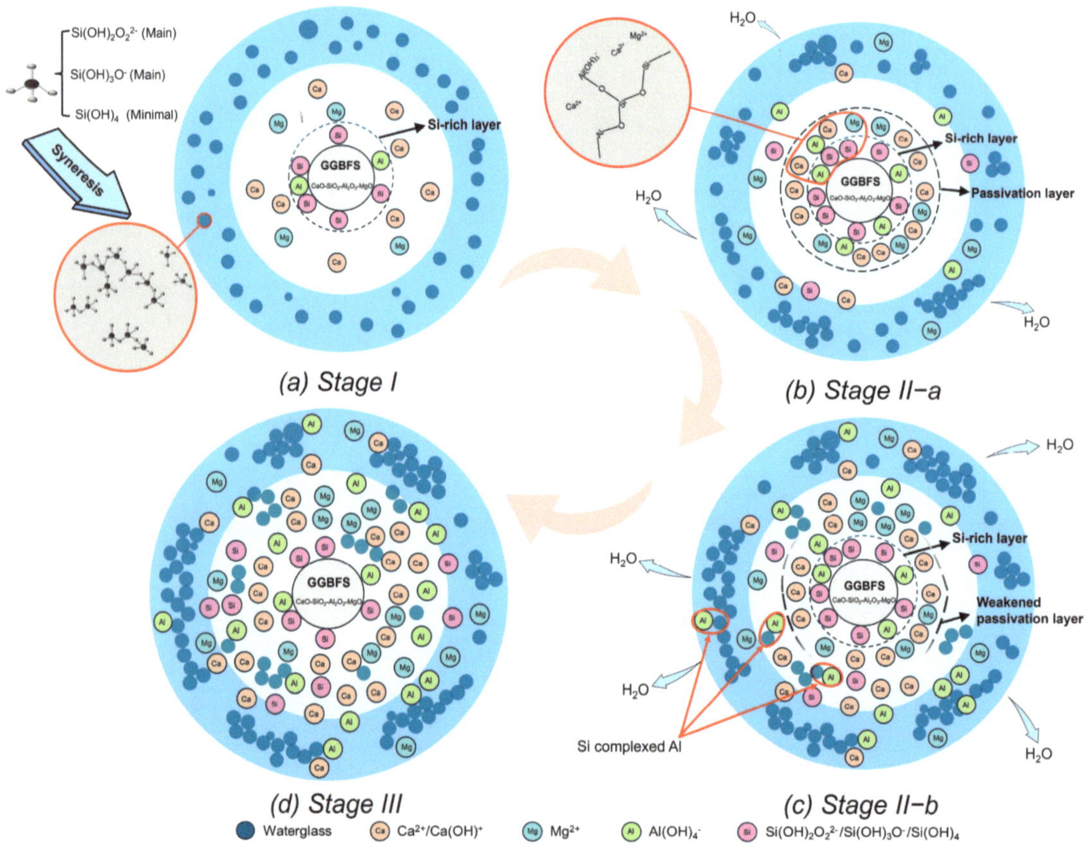

Figure 5. Schematic diagram illustrating the spatial distribution of ions dissolving from GGBFS and silicate structure formation, growth, and aggregation at different stages.

Stage II: As shown in Figure 5b, as the hydration of GGBFS processes, the silicate structure composing the Si-rich layer not only forms a complex with $Al(OH)_4^-$ dissolved in solution (Equation (3)), producing a small amount of water, but also re-adsorbs Ca^{2+}, $Ca(OH)^+$, and Mg^{2+} ions that are dissolved in solution through strong electrostatic attraction. This results in the formation of a passivation layer with a positive surface charge and a negative internal charge. This passivation layer inhibits further hydration, marking the beginning of Stage II-a [31]. As hydration is suppressed, the temperature curve shows a decreasing trend, while the free water content curve exhibits a sharp increase. As previously mentioned, this can be attributed to dissolved ions from GGBFS disrupting the electrostatic repulsion between the original $Si(OH)_3O^-$ and $Si(OH)_2O_2^{2-}$ in the pore solution. This disruption facilitates the aggregation of the silicate structure, leading to dehydration, condensation, and the release of water (Equation (2)) [32]. Water that was originally bound gradually transforms into free water. Coupled with the impeded hydration, the rate of free water production exceeds its consumption. As shown in Figure 5c, with the continuous dissolution of GGBFS, $Si(OH)_3O^-$ and $Si(OH)_2O_2^{2-}$ in pore solution rapidly complex with dissolved ions (e.g., $Al(OH)_4^-$ as seen in Equation (3)), thereby weakening the surface passivation and promoting further dissolution, marking the beginning of Stage II-b [34,35]. A second peak in the hydration temperature rise curve occurs at this stage, indicating the formation of hydration products in large quantities [36]. Notably, while the content of free water continues to increase at this stage, the rate of increase gradually decreases. This is because, in Stage II-b, ions in solution still undergo dehydration condensation reactions, but water consumption accelerates significantly due to the weakening of the passivation layer and the further increase in hydration. This results in a gradual decrease in the net rate of free water production.

Stage III: The hydration temperature rise curve begins to decline, and by 24 h, the temperature of the AAS pastes approaches room temperature. In the free water content curve, no significant changes are observed. At this stage, the reaction rate slows considerably and reaches a stable phase, with hydration predominantly controlled by the diffusion rate, as shown in Figure 5d.

4. Conclusions

This study aims to continuously monitor the early hydration process of AAS pastes using the ^1H low-field NMR technique. The hydration of AAS pastes at their early age was divided into three distinct stages based on changes in free water content and the hydration temperature rise. These findings offer valuable insights into the early reaction mechanisms of AAS. The key conclusions are summarized as follows:

The type of activator significantly affects the change in free water content. AAS pastes activated by waterglass show a substantial increase in free water content at the initial stages, with the increase being more pronounced at higher waterglass moduli. In contrast, AAS pastes activated with sodium hydroxide exhibit only a modest rise in free water content, while activation with sodium sulfate does not result in any increase.

The ions dissolved from the GGBFS in the alkaline environment disrupt the electrostatic repulsion between the ions in the waterglass, promoting further aggregation of the silicate structure. Additionally, these ions may react directly with the waterglass. These processes are likely contributors to the observed increase in free water content.

Based on the observed changes in free water content and the corresponding hydration temperature rise, the early hydration process of WG can be categorized into three stages: dissociation (Stage I), transition (Stage II), and diffusion control (Stage III). The high correlation between the changes in free water content and hydration temperature rise suggests that water may serve as a useful probe, in addition to heat signals, for monitoring the hydration process of fresh AAS pastes.

Author Contributions: Conceptualization, A.G.; methodology, J.Y.; validation, A.G., F.M., T.Z., J.W., Z.S. and J.Y.; writing—original draft preparation, A.G.; writing—review and editing, J.Y.; project administration, Z.S.; funding acquisition, J.Y. and Z.S. All authors have read and agreed to the published version of the manuscript.

Funding: This research was funded by the National Natural Science Foundation of China, Grant No. 52278272, and Grant No. 52408284; the Postdoctoral Fellowship Program of CPSF, Grant No. GZB20240532; the Science and Technology Commission of Shanghai Municipality, Grant No. 23DZ1203500; the Expert Workstation Project of the Science and Technology Department of Yunnan Province, Grant. No. 202105AF150243.

Data Availability Statement: The original contributions presented in the study are include in the article, further inquires can be directed to the corresponding author.

Conflicts of Interest: Author Jiandong Wu is employed by the Shandong Provincial Communications Planning and Design Institute Group Co., Ltd. The remaining authors declare that the research was conducted in the absence of any commercial or financial relationships that could be construed as a potential conflict of interest.

References

1. Dai, X.; Aydin, S.; Yardimci, M.Y.; De Schutter, G. Rheology and structural build-up of sodium silicate-and sodium hydroxide-activated GGBFS mixtures. *Cem. Concr. Compos.* **2022**, *131*, 104570. [CrossRef]
2. Zhuang, K.; Liu, B.; Fang, Y.; Guo, Y.; Shuai, L.; Dong, B.; Long, W.; Xing, F. Is electrodeless resistivity method suitable for monitoring the early-age reaction of Na_2SiO_3-activated slag? Mechanism and application. *Constr. Build. Mater.* **2021**, *272*, 121719. [CrossRef]
3. Fang, Y.; Zhuang, K.; Zheng, D.; Guo, W. The influence of alkali content on the hydration of the slag-based geopolymer: Relationships between resistivity, setting, and strength development. *Polymers* **2023**, *15*, 518. [CrossRef]
4. Yang, J.; Sun, Z.; De Belie, N.; Snoeck, D. Internal curing and its application to alkali-activated materials: A literature review. *Cem. Concr. Compos.* **2023**, *145*, 105360. [CrossRef]
5. Sun, Y.; de Lima, L.M.; Rossi, L.; Jiao, D.; Li, Z.; Ye, G.; De Schutter, G. Interpretation of the early stiffening process in alkali-activated slag pastes. *Cem. Concr. Res.* **2023**, *167*, 107118. [CrossRef]
6. Živica, V. Effects of type and dosage of alkaline activator and temperature on the properties of alkali-activated slag mixtures. *Constr. Build. Mater.* **2007**, *21*, 1463–1469. [CrossRef]
7. Hu, X.; Shi, C.; Zhang, Z.; Hu, Z. Autogenous and drying shrinkage of alkali-activated slag mortars. *J. Am. Ceram. Soc.* **2019**, *102*, 4963–4975. [CrossRef]
8. Fang, Y.; Huang, Y.; Wang, A.; Zhu, C.; Zhuang, K.; Wang, C.; Zheng, D. Integration of monitoring indicators for self-sensing alkali activated cementitious materials: From electrical signals-resistivity to autogenous shrinkage. *Constr. Build. Mater.* **2023**, *409*, 133976. [CrossRef]
9. Fu, Q.; Bu, M.; Zhang, Z.; Xu, W.; Yuan, Q.; Niu, D. Hydration characteristics and microstructure of alkali-activated slag concrete: A review. *Engineering* **2023**, *20*, 162–179. [CrossRef]
10. Haha, M.B.; Lothenbach, B.; Le Saout, G.; Winnefeld, F. Influence of slag chemistry on the hydration of alkali-activated blast-furnace slag—Part I: Effect of MgO. *Cem. Concr. Res.* **2011**, *41*, 955–963. [CrossRef]
11. Haha, M.B.; Lothenbach, B.; Le Saout, G.; Winnefeld, F. Influence of slag chemistry on the hydration of alkali-activated blast-furnace slag—Part II: Effect of Al_2O_3. *Cem. Concr. Res.* **2012**, *42*, 74–83. [CrossRef]
12. She, A.-m.; Yao, W. Research on hydration of cement at early age by proton NMR. *J. Build. Mater* **2010**, *13*, 377–380.
13. Liu, H.; Sun, Z.; Yang, J.; Ji, Y. A novel method for semi-quantitative analysis of hydration degree of cement by 1H low-field NMR. *Cem. Concr. Res.* **2021**, *141*, 106329. [CrossRef]
14. Yang, Y.; Liu, J.; Liu, L.; Li, J.; Liu, Q.; Chen, Z.; Shi, C. Quantifying the water saturation degree of cement-based materials by hydrogen nuclear magnetic resonance (^{1}H NMR). *Constr. Build. Mater.* **2024**, *438*, 137340. [CrossRef]
15. Yang, J.; Snoeck, D.; De Belie, N.; Sun, Z. Comparison of liquid absorption-release of superabsorbent polymers in alkali-activated slag and Portland cement systems: An NMR study combined with additional methods. *Cem. Concr. Res.* **2021**, *142*, 106369. [CrossRef]
16. Yang, J.; Sun, Z.; Zhao, Y.; Ji, Y.; Li, B. The water absorption-release of superabsorbent polymers in fresh cement paste: An NMR study. *J. Adv. Concr. Technol.* **2020**, *18*, 139–145. [CrossRef]
17. Liang, G.; Yao, W.; She, A. New insights into the early-age reaction kinetics of metakaolin geopolymer by ^{1}H low-field NMR and isothermal calorimetry. *Cem. Concr. Compos.* **2023**, *137*, 104932. [CrossRef]
18. Jiang, D.; Zhang, Z.; Shi, C. Evolution of silicate species from waterglass in waterglass activated slag (WAS) pastes at early age. *Cem. Concr. Res.* **2024**, *180*, 107513. [CrossRef]
19. Zhang, D.-W.; Zhao, K.-f.; Wang, D.-m.; Li, H. Relationship of amorphous gel-microstructure-elastoviscosity properties of alkali-activated materials fresh pastes with different Ms waterglass. *Constr. Build. Mater.* **2021**, *287*, 123023. [CrossRef]
20. Wang, S.-D.; Scrivener, K.L. Hydration products of alkali activated slag cement. *Cem. Concr. Res.* **1995**, *25*, 561–571. [CrossRef]
21. Haha, M.B.; Le Saout, G.; Winnefeld, F.; Lothenbach, B. Influence of activator type on hydration kinetics, hydrate assemblage and microstructural development of alkali activated blast-furnace slags. *Cem. Concr. Res.* **2011**, *41*, 301–310. [CrossRef]
22. Gebregziabiher, B.S.; Thomas, R.; Peethamparan, S. Very early-age reaction kinetics and microstructural development in alkali-activated slag. *Cem. Concr. Compos.* **2015**, *55*, 91–102. [CrossRef]
23. Cao, R.; Zhang, S.; Banthia, N.; Zhang, Y.; Zhang, Z. Interpreting the early-age reaction process of alkali-activated slag by using combined embedded ultrasonic measurement, thermal analysis, XRD, FTIR and SEM. *Compos. Part B Eng.* **2020**, *186*, 107840. [CrossRef]
24. Yang, J.; Sun, Z.; Li, B.; Ji, Y.; Hu, K. Measuring volume change of alkali-activated slag pastes in early stage by using helium pycnometry. *J. Mater. Civ. Eng.* **2019**, *31*, 06019011. [CrossRef]

25. *GB/T 12959-2008;* Test Methods for Heat of Hydration of Cement. Quality Supervision Inspection and Quarantine of the People's Republic of China and Standardizetion Administration of People's Republic China: Beijing, China, 2008. (In Chinese)
26. Bernal, S.A.; Provis, J.L.; Rose, V.; De Gutierrez, R.M. Evolution of binder structure in sodium silicate-activated slag-metakaolin blends. *Cem. Concr. Compos.* **2011**, *33*, 46–54. [CrossRef]
27. Shi, C.; Roy, D.; Krivenko, P. *Alkali-Activated Cements and Concretes;* CRC press: Boca Raton, FL, USA, 2003.
28. Yang, K.; White, C.E. Modeling of aqueous species interaction energies prior to nucleation in cement-based gel systems. *Cem. Concr. Res.* **2021**, *139*, 106266. [CrossRef]
29. Šefčík, J.; McCormick, A.V. Thermochemistry of aqueous silicate solution precursors to ceramics. *AIChE J.* **1997**, *43*, 2773–2784. [CrossRef]
30. Kim, T.; Olek, J.; Jeong, H. Alkali–silica reaction: Kinetics of chemistry of pore solution and calcium hydroxide content in cementitious system. *Cem. Concr. Res.* **2015**, *71*, 36–45. [CrossRef]
31. Jia, R.; Wang, Q.; Luo, T. Understanding the workability of alkali-activated phosphorus slag pastes: Effects of alkali dose and silicate modulus on early-age hydration reactions. *Cem. Concr. Compos.* **2022**, *133*, 104649. [CrossRef]
32. Matinfar, M.; Nychka, J.A. A review of sodium silicate solutions: Structure, gelation, and syneresis. *Adv. Colloid Interface Sci.* **2023**, *322*, 103036. [CrossRef]
33. Yu, S.; He, J.; Sang, G.; Yang, S.; Liu, G. Study on hydration process of alkali-activated slag cement activated by weakly alkaline components. *Constr. Build. Mater.* **2024**, *413*, 134716. [CrossRef]
34. Huanhai, Z.; Xuequan, W.; Zhongzi, X.; Mingshu, T. Kinetic study on hydration of alkali-activated slag. *Cem. Concr. Res.* **1993**, *23*, 1253–1258. [CrossRef]
35. Aliabdo, A.A.; Abd Elmoaty, M.; Emam, M.A. Factors affecting the mechanical properties of alkali activated ground granulated blast furnace slag concrete. *Constr. Build. Mater.* **2019**, *197*, 339–355. [CrossRef]
36. Fernández-Jiménez, A.; Puertas, F. Alkali-activated slag cements: Kinetic studies. *Cem. Concr. Res.* **1997**, *27*, 359–368. [CrossRef]

MDPI AG
Grosspeteranlage 5
4052 Basel
Switzerland
Tel.: +41 61 683 77 34

Buildings Editorial Office
E-mail: buildings@mdpi.com
www.mdpi.com/journal/buildings